WHAT SEEMS TO BE THE TRUTH TODAY,

YOU WILL ABANDON

FOR A HIGHER TRUTH OF TOMORROW.

The Sequence Of Events

Documenting When Light Was In It's Infancy. / Ending Yesterday.
For Readers Over The Age Of Thirty Only

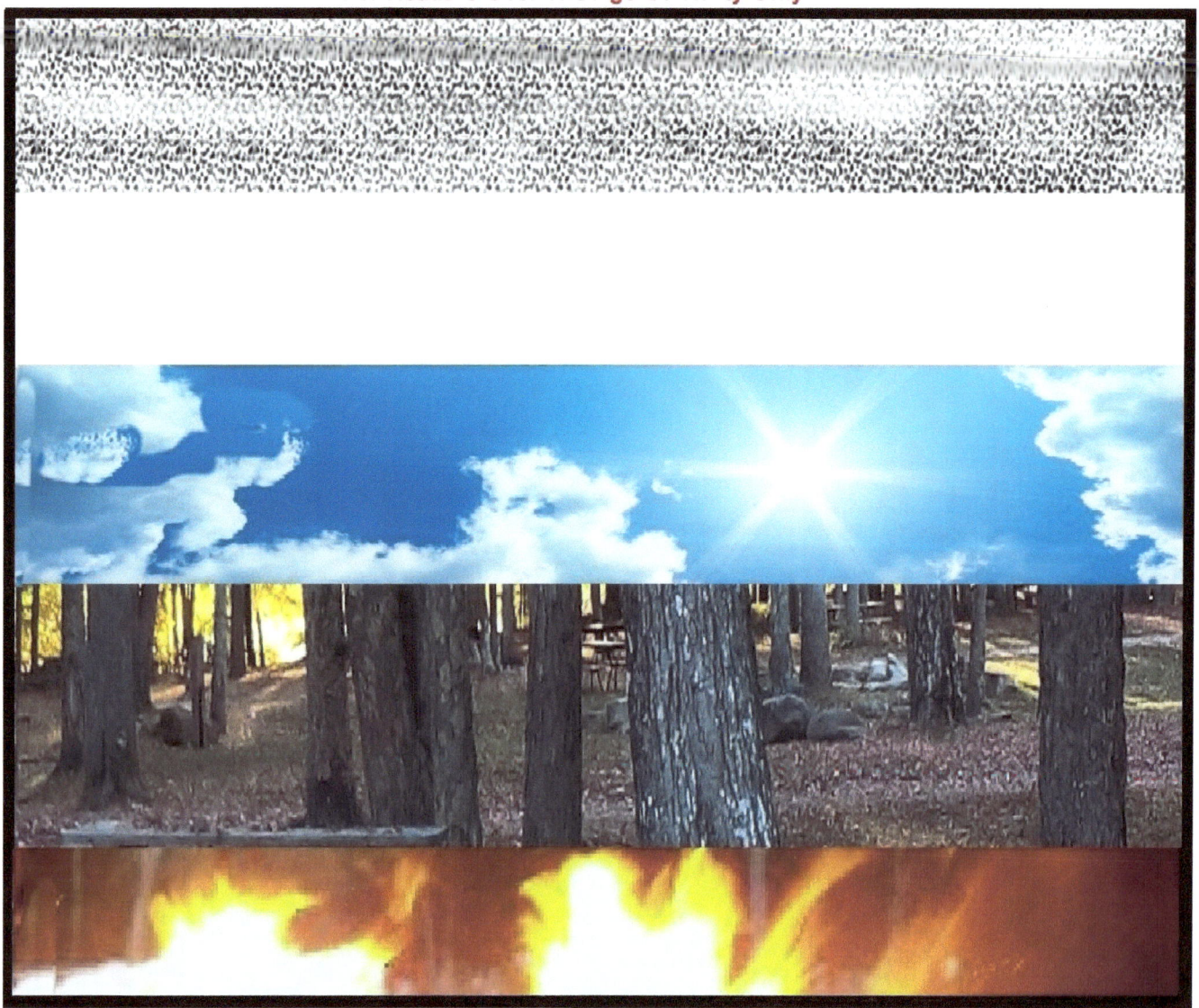

This Is A Science / Religion Documentry

Be Advised, This Is A Science Manuscript

If You Do Not Read Page One

You Will Not Understand Page Two

At one time, there was once nothing.
There was no heaven, no god, no hell, and no earth. There was nothing that could be understood or comprehended by man. There was no substance that could be named or described. There were only bits of frozen, nameless parts, scattered in a world of total darkness. The total size of this world that was made of darkness would be past the wildest imagination of any man.

This Documentary Is A Rendition
Of
The Writings Of The Finest Minds That Ever Existed
In Regard To Science And Religion.

The Author Leo Spaziano Hopes That You Profit By Examining;

The Sequence Of Events

CONTENTS

Introduction……………………………………………………Page 7

Who Is The Author?………………………………………Page 8

Glossary Of Terms…………………………………………Page 9

The Birth Of The First Light……………………………Page 14

The First Molecules ………………………………………Page 18

The First Living Beings……………………………… Page 22

The First Human That Evolved…………………… Page 44

The Necessary Birth Of Hell……………………………Page 48

Who & What Is The Primary One?………………Page 74

Hell…Q & A…………………………………………………Page 83

CONTENTS II

Do Animals Have Souls? ………………………………Page 116

Dreams………………………………………………………Page 120

Who Is The First Devil?………………………………Page 122

Foreclosure………………………………………………Page 126

Blessings Of All Faiths…………………………………Page 132

Dedication And Credit Page…………………………Page 136

Biography Of The Author……………………………Page 137

Letter of Thanks…………………………………………Page 139

Copyright Page……………………………………………Page 140

Introduction To The Sequence Of Events

The Sequence Of Events will document when light was in its infancy. This book will describe, a believable understanding and a description of the first moments as it has never been explained before.

The only way to comprehend the un-comprehendible is by measuring things in the spirit rather than measuring things in the mind or the flesh. The spirit is strong and it is limitless, but the flesh and the mind are highly limited and weak.

The author will not use his own wisdom to explain these events since human wisdom is highly limited. Since no person has a monopoly on wisdom, the author will present, prewritten and ancient records. Those records are so scientifically believable and supernaturally enriching; that they can give way to explain the unexplainable.

This documentary was not just thrown together in twenty five minutes, in some dark basement. The ancient writings were scrutinized and heavily studied since 1986. The writings were edited for simplicity and for speed, in hopes to seduce the reader to become morally prosperous and financially enriched.

Who Is The Author

The author claims to be ordinary in intellectuality and skill. He claims to be the collector of rare scientific and religious documents. He suffers from severe anxiety, on the subject of who we are in the cosmos.

By reason of the spirit instead of intellectuality, he tries to grant peace to those who seek true knowledge and true wisdom. He attempts to give aid to those who find themselves lost in the foulness, and unanswered questions of science, mathematical functions and life as we know it today.

Glossary Of Terms

Definition Of Life
Man is made of a hand full of minerals and some water. Basically, man is made of dirt and moisture and yet he thinks of himself as the absolute ruler of the earth and prides himself as the absolute intellectual being.

Something that always points toward the sun for light energy is, the leaves on a tree and the petals on flowers. They always move and follow the sun, as it skips across the sky. This motion takes intellectuality, and it takes thinking power. But no one knows where a shrubs brain is located or if each leaf or if each petal has its own brain.

Many varieties of octopus, can land on the floor of the ocean and immediately change the color of their body to match the color of the ocean floor. They camouflage themselves for protection. It is in this way; they can blend in and not be victimized by predators. This takes an enormous amount of thought and intellectuality. This color process is a total mystery to man.

The octopus and the movement of shrubs are a total mystery to mankind. It could be presumed that man's understanding of life is, no understanding at all. What life actually is, escapes man's intellectuality. The reality is, that man may have overlooked the possibility that life could be constructed of light or protons instead of sand and water.

If there is a possible chance that life could be composed of light, this could explain the reports made by many honest people who have claimed to see unusual things. They may have seen things like, semi transparent images (ghosts) or they may have claimed to see shiny vehicles of impeccable motion in the sky. This type of life form could live forever and never die.

If this form of life is possible, extreme distance travel and extreme time duration have no value at all, since this life form has forever to go anyplace and it has forever to do anything. One might argue, how can light live ?   Man is made of dirt and moisture. How can man live, and yet man lives.

Hence,
Man does not understand all forms of life because his mind is too puny. This does not mean that a life does not exist at all, just because man can not accept or understand what life is.
The definition of life is, "an infinite variable".
This is what life means.

This moth has excellent vision and it can see in color. It often sleeps next to leaves that have the same color it has, in order to avoid being eaten by birds. Man can not make a color camera as small as the size of a moth's eye.

How Far is Far ?
If your automobile could travel 186,000 miles per second, it would take you 13 billion years to get to the farthest known group of stars. These stars are 75,980,000,000,000,000,000,000,000,000 miles away.
This is what Far means

What Time Is It ?
Since it is 13 billion years away from the farthest known group of stars, one could imagine that we live in the year 13 billion, East Greenwich Time.
This is what Time means

How Big is Big
Astronomers believe that;
If you just counted the pounds that the earth weighs, one pound at a time, it would take you and 26 trillion generations of your family, to count up to only 1% of the total weight of the earth.

Measuring the weight of the earth might be beyond our comprehension. It is impossible for us to count out loud up to two billion, even if we lived to be one hundred years old and if we counted all day long.

The Earth's Sun and its other surrounding planets weigh 6 million times more than the earth itself. You would have to live forever to experience the cosmos. **This is what big means**.

So this is the best description of life, distance, time and size.

The Birth Of The First Light
At one time, there was once nothing. There was no heaven, no god, no hell, and no earth. There was nothing that could be understood or comprehended by man. There was no substance that could be named or described. There were only bits of frozen, nameless parts, scattered in a world of total darkness. The total size of this world of darkness would be past the wildest imagination of any man.

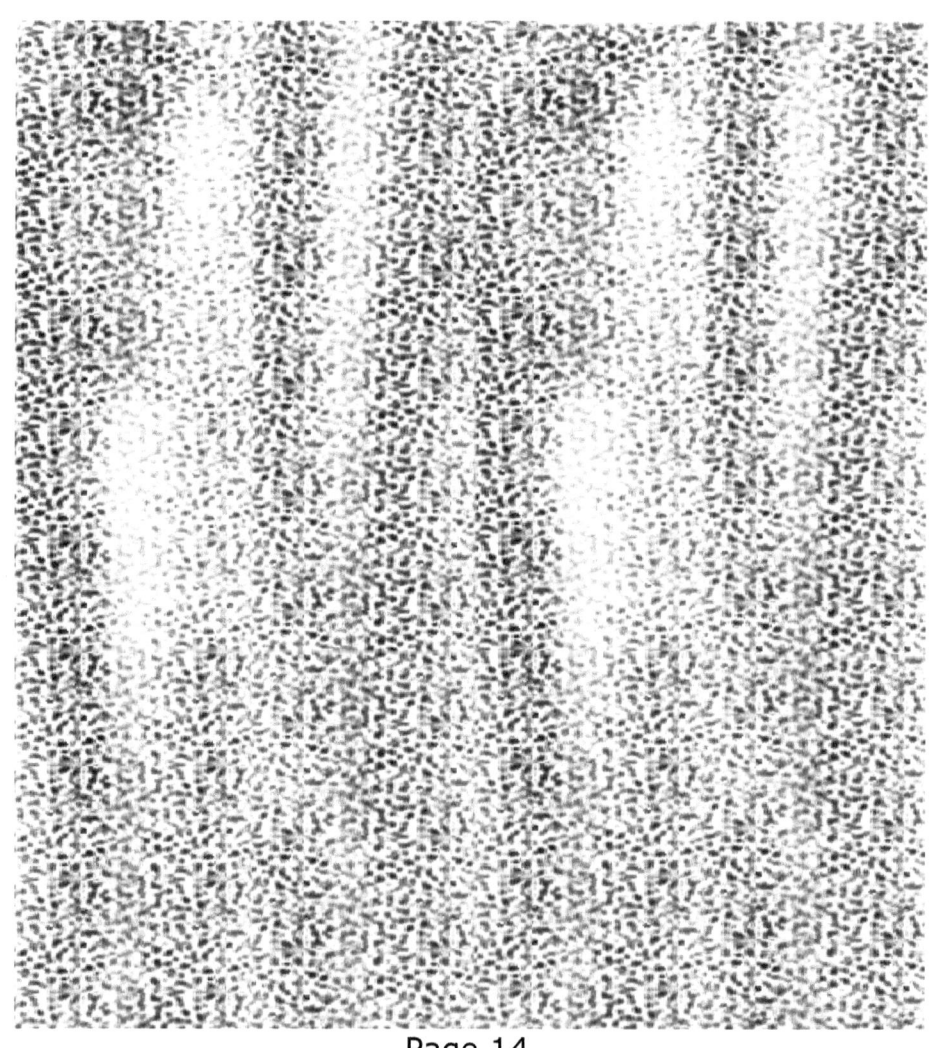

These bits of nameless parts began to cling to each other. They began to grow in size as they grouped together. As they developed a heavy weight, they also began to develop a stronger gravitational pull, or a gravitational attraction.

The gravitational pull of these bundles or parts had developed enough gravity to attract themselves to formations that were similar to themselves in type. As the parts collided and combined with each other, they grew in their size, in their weight, and their gravitational pull increased. This growth continued until the combined weight of all these items, developed an immense amount of weight, heat and gravity.

As the crushing gravity of this mass grew to be too large, heat began to form in the center of those bundles of particles. Eventually, this growth was unimaginable in size. The heat was so much, that an explosion took place. This explosion was so big that it would make the big bang theory look like a match stick. A better description of this explosion would be "the biggest bang theory"

This explosion gave birth to the first light. This first light burst that took place, generated a wide band of different groups of light that would escape the resolution of a man's eye by leaps and bounds. The initial emission of this light, which also contained invisible gamma light, could easily penetrate and totally dissolve lead.

This explosion of light splintered off into fragmented and tangent, firebrands better know as stars. Some of the stars, suns, or firebrands were enormous. The largest sun is two thousand, six hundred times the size of the earth's sun. The smallest known sun, tangent, firebrand, or star is only about one tenth of the size of the earth's sun.

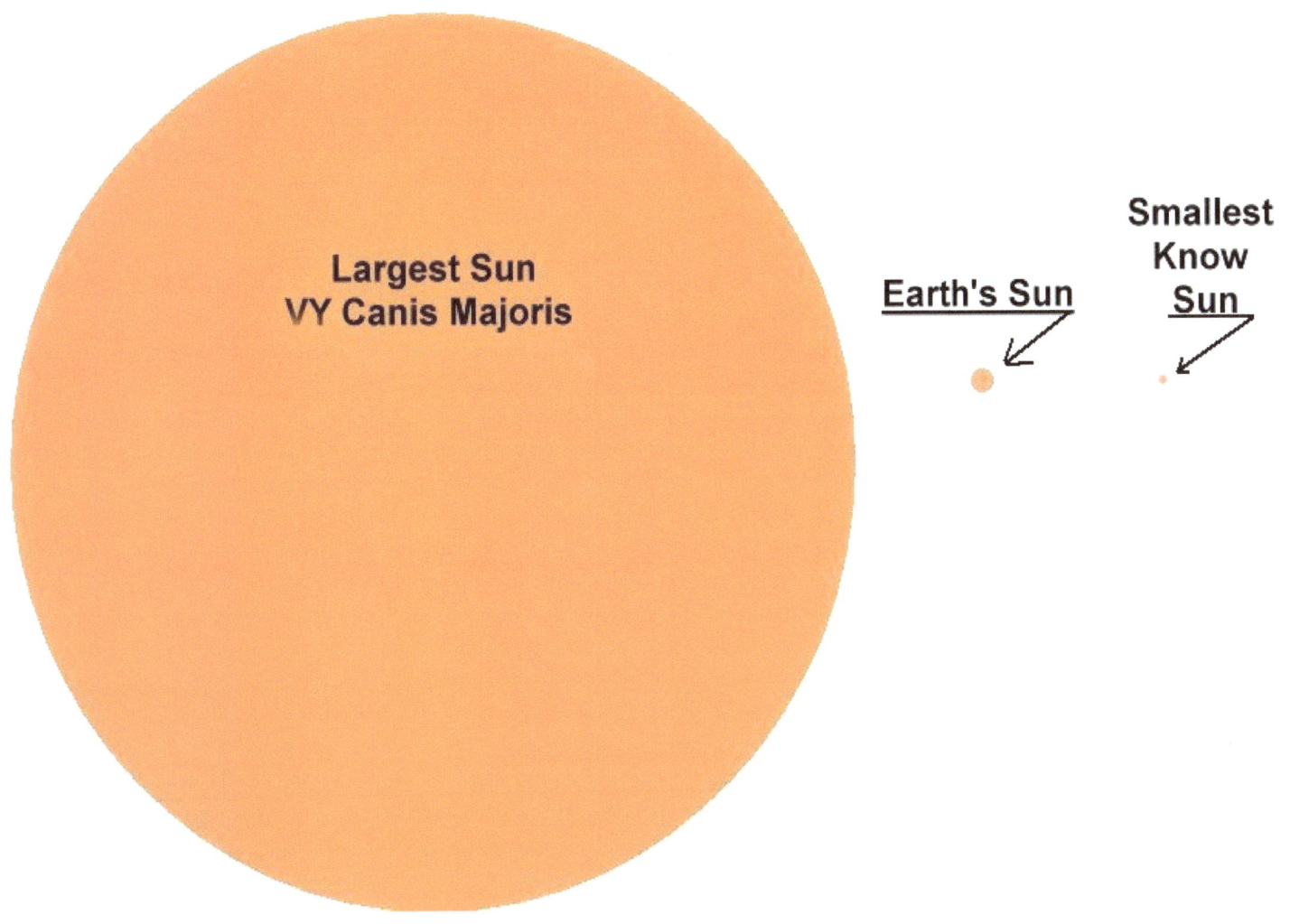

The Formation Of The First Substance And The First Molecules

After this explosion took place, these light emissions slowed down in speed a little at a time, due to simple friction. As the light or protons slowed down in speed, they did not glow with light anymore. As a result of slowing down, the particles of light (protons) turned back into a solid mass.

So this cycle of substance made of light, went something along these lines:
1) The unnamed substances grouped together into lumps of mass.
2) The heavy lumps of mass exploded and formed light.
3) The light particles slowed down in speed and failed to give light anymore.

The particles of light that slowed down and cooled off the fastest, turned into some material that weighed very little. The particles that took longer to slow down or cool off weighed very much. The heavy particles were metallic and the lighter particles were gaseous.

No matter how many times this concept of light turning into substance takes place, the result is always the same. It seems as though, light can turn into only ninety two distinct formations of material substance. The lightest in weight is called hydrogen gas. Examples of the more heavy material substance made out of this light is called lead, mercury and so on.

Liar
Liar
Pants On Fire

A famous astronomer comes forward and opposes the number 92. Who...... does this collector of rare documents thinks he is. He was once a janitor at the YMCA in 1973. And now......this son of a nobody wants us to believe that the whole cosmos is made of only 92 parts.

Why... this author has hardly put his foot on the moon. He hasn't visited the closest stars. In past scripts, in which the author has published, the author misspelled the word "astronomer". He can't even spell. The author was once arrested for stealing Home Box Office Television.

And now....... the author wants to sell us this ridicules story about the whole cosmos being made out of only 92 parts.

The Author rebuts the famous astronomer;
Everything the famous astronomer has said is true, however, the insulting astronomer cannot explain to us why his own bowel movement is always brown regardless of the color of the food he puts in his mouth. The astronomer sometimes goes to his place of employment with his necktie on crooked.

No matter what color food you eat, the outcome is always the same color of brown. How odd is this?

There is a thin line in regard to this process of light turning into substance. It seems as though, there is a middle-ground of the process in which the substance is half light and half solid material.

Some of the substances in this mid-state range causes noise on radio receivers and is termed radioactive. This noise interferes with radio communication. Any substance that makes noise and interferes with radio communication is termed radioactive.

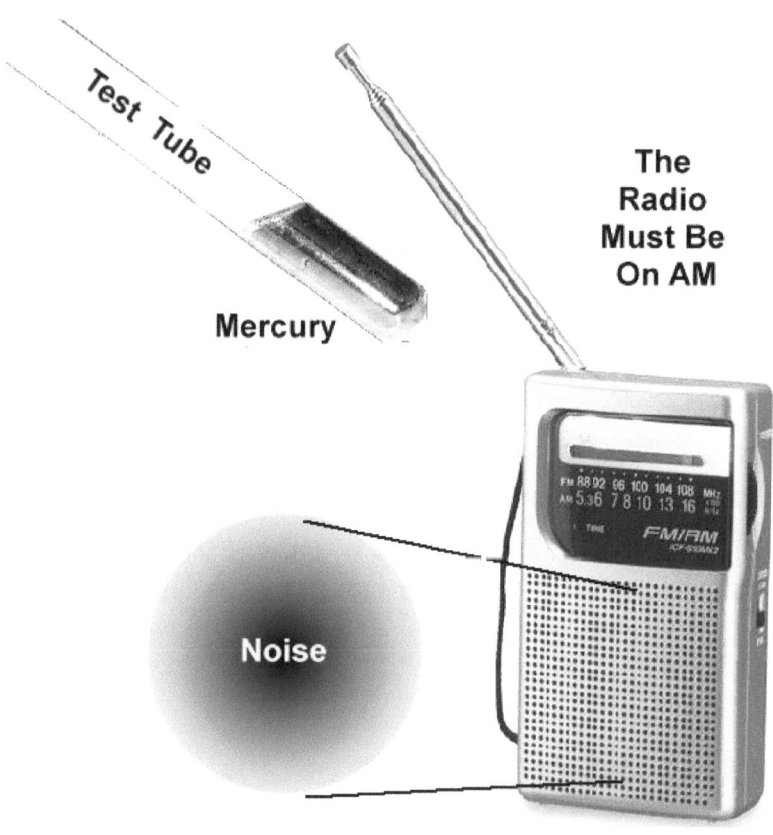

The Invention Of Einsteinium
Pronounced ein•stein•i•um
Einsteinium is a mid-state, radioactive, <u>man made substance</u> that is always produced artificially and does not exist naturally in the cosmos. It was named in honor of Albert Einstein.

This is a radioactive metal and was discovered in 1953. It was buried in the waste that was left behind from the first hydrogen bomb explosion. It is usually produced by the collision of other natural, radioactive, mid-state substance or element.

Einsteinium is a substance that does not occur naturally on Earth. It can only be created artificially. As of today, there are 24 synthetic, radioactive substances.

The First Living Beings
The definition of life is outlined on page nine of this manuscript.

To recall;
- Man, is made of a hand full of minerals and of some water.
- Man does not understand the mechanics of why flowers and leaves point toward the sun for light energy.
- Man does not understand the concept of the process of the body of an octopus changing its color, in which allows the octopus to blend in with its surroundings.

Since man can not understand what life actually is, there is a great possibility that life constructed of light or protons, instead of sand and water, may be a scientific reality. Man's lack of intellectuality does not limit the growth or the intellectuality of another life form.

In the same way that the accumulation of dirt and water is the primary formula to make mankind and trees, one may conclude that the accumulation of light connecting to more light may give way to an intellectual life form made of light itself.

These Are Approximate Writings Of The Finest Authorities, Regarding Photonic Life Forms

Rita Rizzo in Canton, Ohio..Demanded To Be Copied

Ivan Dragicevic ....Waived Copyright Without Prejudice

Farmer Bartholomew of Italy.................After Great Thought Waived Copyright

Mirjana Soldo..Waived Copyright Without Prejudice

Ms Magdalene of Magdala..Demanded To Be Copied

John Wilson of Boston...Demanded To Be Copied

Ivanka Elez ....Waived Copyright Without Prejudice

Marija Lunetti...Waived Copyright Without Prejudice

Jakov Colo ....Waived Copyright Without Prejudice

James of Bethsaida....................................Allow Copyright With Reluctance

Vicka Ivanković ....Waived Copyright Without Prejudice

Thomas of Galilee............................ After Great Thought Waived Copyright

Philip of Bethsaida..Demanded To Be Copied

Benedict Groeschel ....Demanded To Be Copied

The First Light Based Life Forms

It is thought of that eons before the earth came about, that random luck and evolution would mandate the creation of living forms of life, that were made of light particles or protons. There is this theory of the evolution of life that was formed and made up of several combinations of random particles of light or photons.

This theory of a light based life-forms that evolved, was no walk in the park. Just as in the theory of the evolution of man, it took many trials and many errors in these formations in order for the life-forms to become perfected.

Once, when the light based infant finally came about, it began a self growth in which there was no mechanical problem that it would not be able to overcome. In fact, there was only one rule. That rule was "Do Not Defy Logic"

A series of generations of these creatures made so many wise decisions, that it raised their life expectancy to an unlimited lifespan. A scientific explosion took place where there were so many generations of good decisions made by these creatures, that immortality was quickly maintained.

Many of these new generations of light, were wiped out of their own existence due to making poor choices. They defied logic and perished. Only those who made wise decisions remained.

The First Argument and Confrontation

These beings had a monopoly on manufacturing. As for instance, they could present themselves to others in any fashion they wanted to. They could easily change there appearance. Because the life forms were made out of light, they could imitate anything including the stars in the cosmos.

A single one of these life forms could easily imitate what would look like a whole army of creatures just like itself to seemingly appear to someone else. And then, the downfall came about. They created deceit. They argued among each other about who was the greatest of their group.

One life form tricked another life form into believing it was at war with a hostile foe in which the foe had no real formation or had no real existence. So, the life form that was tricked into thinking it was at war, had exhausted itself fighting in a war that never even existed and it died off as a result of exhaustion. The whole war was a lie.

As they spun themselves into death, only one life remained intact. All the other light based life forms, killed off one another. The weapon of choice was deceit. At this time, it was impossible to produce a natural light based, life form anymore. The dynamics of the original explosion of the unknown material or the so called, the biggest bang, no longer naturally produced life.

The last surviving creature was alone. The creature was unique. It lived in a world where there was only random light and this light was slowing down in speed. This light was turning into a useless, senseless, stagnant substance.

The Invention Of Yourself

This light based creature was alone for countless amounts of eons. It wandered around through space and places of emptiness. The creature did this for so long that it forgot what its name was since it hasn't been addressed by anyone in a multitude of time.

It would usually be difficult for one to forget their own name, but the loneliness and the new experiences of the cosmos wiped out its need to be called anything. Since this creature had no friends, no mother and no father, it had no name.

The light based creature was somewhat sexless. It had the kindness that woman might have and at the same time it could be as strong as a man could be. So, as it went about in time, it thought of itself as Mother and Father with no offspring. The lonely light based, photonic, creature, desired company.

Consider that a human can pick up snow and make a snowball. This unique creature could pick up light as it passed by, and make a formation of life which would be in the same likeness and formation as itself.

A serious problem in which had to be pondered upon was the deceit, greed and hate that killed off those creatures he once knew. He remembered very well <u>those moments</u> in which they died off and they turned from the form of a beautiful light into a substance of static value or no real value at all.

Out of loneliness and desperation, it grasped light as it passed by, and it reformed that light. It made a light based life form exactly and identical to its own likeness and existence. This new life form was an exact duplicate of itself. This would guaranty itself that there would be no conflict, no arguing and no chance of a war.

One may say, how stupid is this. In order to understand this self invention, think of yourself as to be looking into a mirror as you might be combing your hair and talking to yourself out loud. Or you might be the type to sing when there is no one around to sing to.

Or perhaps you can imagine hollering to a neighbor or hollering at a bad motorist that you know cannot hear you. So in this way, you temporally invent yourself, as you keep yourself company since no one is listening.

So it tried to have a better existence. It decided to name this life form. This was a very difficult thing since the life form was an exact duplicate of itself. Through the existence of this new life form, it found some salvation. It gave itself good advice through this new creation.

It also aided the behavior of this new life form and gave it knowledge. So it decided that this new life form was not only its genderless son, but it was also its genderless father at the same time. The life form mirrored and returned its own thoughts back to itself.

The Creation Of The First Indestructible Soul

Countless amounts of eons went by as this unique, light based, life form, and its father/son, co-existed. The cosmos was still being manufactured by the explosion of the nameless parts. The world seemed to be dull, uninteresting, mundane, monotonous, and empty.

This son suggested to its father; let us make new bodies of light, bodies just like ours. If we make these creations, the world will not seem to be monotonous, dull, and uninteresting. The father rejected this idea as he very clearly remembered the horrible past chaos and wars.

As this idea was being pondered, the primary one, decided to make a totally new being. This being would be flawless and indestructible in its natural state. He made the new life form out or perpetual light that could never be destroyed. Neither could it ever become permanently corrupt.

The primary one, did not want this new perpetual, light based, life form to suffer from dullness or monotony. So, he granted this perpetual light based life form, a gift. This gift was very temporary and it was totally retractable.

This temporary gift was called free will. Hence the first souls were made of indestructible, perpetual, light with a temporary gift of free will. This new life form would be indestructible, and could be redeemable regardless of the paths it chose to take.

The Invention Of The Soul
A problem arose with the ones of perfection. The bliss-ness and jubilation were well perfected but it had a flaw. The flaw was that, basic physics mandated boredom to exist within it. They found a need for more fulfillment, internal satisfaction and a need to refine what they thought was perfection.

So, they created a life form that was made of several facets, which could make use of silliness and folly. Silliness would be somewhat of an electable selection by the life form itself. Through a selection of paths, those who mastered mortality would find a greater reason for their own self-existence.

The new life form was comprised of an assembly of four parts.

The Soul
The first part was a self-repairable, indestructible and incorruptible substance, which was made with the makeup and the wisdom of the perfected ones who mastered immortality. This part can be referred to as a soul. All souls are about the same in size. The soul is a true perpetual life form.

The Spirit
The second part of this new life form is a device that could manufacture anxiety. This anxiety may navigate the soul and take it for a proverbial ride. Its purpose is to make proposals to the soul in hopes to make the soul feel more versatile. The second part will be referred to as the spirit. All spirits are about the same in size, in terms of desire. The spirit is want for exploration.

The Mind
The third part is a complex device that sits between the soul and the spirit. It acts as an informational, transportation system between the soul and the spirit. This conduit might be thought of as a massive national telephone wiring system, which could connect many messages.

This system could allow vast amounts of information to be transferred from the spirit to the soul. This part is referred to as the mind. The mind is a transceiver or a two way instrument.

The Body
The fourth part of the new life form is nothing more than a case or a box to hold or to unify the soul, the spirit and the mind. Every case is unique. Each body varies in size.

The life form container, or the body, is totally inconsequential and will never satisfy man's definition of what life formations actually include. A tree, an insect, a microorganism, an elephant or a human could be a good example of the case or a box style. Hence, all the new life forms will be comprised of a soul, a spirit, a mind and box or a body to put them in.

After the life form that was created runs its life course, the case, box, flesh, bark or body, becomes totally useless to the ones who mastered immortality. It becomes quite disposable. The conduit or the mind is also disposable and no longer has any use. As a result of this disconnection, the soul is redeemable.

The spirit, anxiety, or the want, separates from the soul and the spirit merges itself into nearby boxes, cases or bodies that it was next to. This is done by reason of closeness or by proxy and may exist for a long period of time.

An example would be. "My neighbor chops wood and keeps himself warm. I shall do as he did. I shall chop wood and be warm as he was." Hence his spirit lives in me." Years ago, a man invented the first automobile. I shall make a more efficient automobile. The spirit and the memory of the man who made the first automobile now lives inside of me.

And now the question is, what happens to the soul which is intended to be an indestructible and an incorruptible substance? There is no record of a soul being destroyed, although the mention of this action is suggested.

Since the soul is mostly indestructible, it needs a new case or body to be stored in. The older body decayed or died off. As a result of this body dying, the ones who mastered immortality needed to redirect the location of the soul.

The soul wants to rest after it has been taken on a journey by means of the spirit, the mind and the body that it has recently been stripped away from. Those who made the soul, now feel a desire to measure it, or weigh it up, to see if the soul has maintained its original purity.

The Elite Ones argued with one another about where the soul can be placed. The powers of creation have opposing solutions to the problem. Internal chaos takes place among the powers. So with all their accomplishments, there is still discomfort and tenseness among these perfect ones.

They would like to dispose of the soul, but they had trouble dealing with disposing of something that is made of indestructible, perpetual light. They also loved what they made with their own hands and they still yearned for it to live. By mastering immortality, they invented a new problem.

Meantime the soul begs for a place to rest and forever finds itself in court, regardless of how many times it has been convicted or pardoned. The soul is in agony and tries to negotiate communications with other souls by way of dreams. It seeks a body the same way a human wishes to live and not die.

So the soul suffers from unrest. The soul suffers so much unrest that it begins to accuse everyone. A split takes place between all the souls, all the spirits, and all the bodies. There is division in all the powers in which strife is the outcome.

Since this is a universal problem in the cosmos, there is no safe place for the soul to hide. This is the reason why the soul is indestructible and the cases, boxes or bodies that house the soul, suffer as well.

Whether one cares to believe in hell, heaven or any other place, the soul finds itself, in an unsolvable predicament. It seems there is no real solution even if the soul was all understanding. The soul becomes confused and finds itself in a cyclical, irresolvable problem, in which it begins to seek refuge by reason of avoidance.

Avoidance is quite a bit of a juggling act. When the soul begins to find peace, other life forms go looking for it. They disturb it and agitate it. The sub author's conclusion is, there is no solving this problem. Total joy is impossible to accomplish without having total discomfort existing in the first place.

You must be hungry first in order to feel satisfied with food.
You must be cold first in order to feel the comfort of heat.
You must feel pain in order to feel happiness or relieve.
You must die in order to feel real life.

May those who are above us, who have mastered immortality, remedy this process of "the hot / cold syndrome" They suffer just as much as we do and the sweet souls that have escaped our vision, suffer as well. I speak of the sweet souls that their spirit sits in our bodies as we think of them and as we wish they were alive.

What ever concept you produce is flawed except for the concept in which there is no concept. Eat, sleep and be merry.

The First Servants

The primary one and his son were still suffering from acute anxiety, loneliness and emptiness. Again, the son made a proposal to make an unlimited amount of helpers with indestructible, perpetual light based, life forms that would be prone to promote prosperity without corruption by giving each one of them a reconcilable soul.

The two of them also thought that the loneliness would be totally eradicated if each of these new forms to be made, could help them in the production of this quest for prosperity. So the primary one took a hand full of fire and produced seven superintendents to help them carry out this mission.

They were all created by the will of the son and with the consent of the father. The seven will be the rod bearers (lectors or minor officials) of the primary one. The primary one named each of the seven superintendents.

Then, with a handful of fire, he made the first one who was the most capable and the most unique. His name was Satanael
Next was Michael, the chief captain.

Then, with another handful of fire, Gabriel was formed third.

Uriel was fourth to be formed. Raphael was fifth to be formed.

Nathanael was sixth, then Zathael, and 6,000 other servants of who I cannot tell you their names.

The primary one kept a written book of remembrance. This book housed all of the names of all these creations. All of these bodies, souls, and spirits, in which he gave birth to, were written into his book of life.

His letters are not vowels, nor are they consonants as seen in language. One might read them and think of something foolish, but rather they are letters of the truth. They alone speak to those who know them.

Each letter is a complete thought, like a complete book. These letters were written for the groups, in order that by the means of his letters, they should know the primary one with a greater understanding.

Who Exactly Is The Primary One
His Name Is;

- In some Slovak countries he is called Bog
- For the god in Hinduism he is called Krishna
- In some Arabic countries, he is called Ala
- In some Eastern countries he is called Jehovah.
- In some Western countries he is called Christ,
- Akal Purakh, meaning timeless One.
- Nirankar, meaning formless One.

The list of The Primary One's names is nine hundred in length and the list has been shortened for simplicity. The sub author personally prefers to refer to the primary one as being called, The One Who Is All Powerful, All Loving And All Understanding.

The Primary One is as old as the light around him, but the light around him has not acquired or exceeded his knowledge. He is immortal, eternal, and has no birth that a man's conscience can grasp. He has no known father and he has no known beginning that man can understand.

Whoever has a name, is the creation of another. He is unnamable. He has no human form. He has a strange appearance that surpasses all things in the universe.

Who Is The Primary One ?

For all practical purposes and for all practical measurements that man can make, all things that exist around him, belong to him. Who will be able to utter a name for him, except for him alone to which the name actually belongs to. He is un-produced. He gave a name to himself alone, since he alone sees himself. He alone has the power to give himself a name.

His constant companion and his philosophy were logic, and wise thinking. Wisdom, from the very first moment, was destined to unionize itself with him. He was First Divinity and First Kingdom.

First Heavens / First Kingdom

The first groups of bodies, with incorruptible souls of free will were finally established. He created a great group of seven deities or god-like associates, for his own use, majesty and pleasure. They were given great authority. He created gods, angels, and archangels, without number for his followers.

It is from that Light and from that Spirit, which is made of or comprised of wisdom, philosophy and logic, who is his constant companion. From this, he originated holiness and kingdom. He has a unique mind and a unique thought. His thought has power, is considering, reflecting, and has rationality.

In respect to these that he made, they are all indeed equal. In respect to power, they have this difference. Their deference is similar to the difference between father and son.

There was a structure that represents these calculations, defined as "the sequences of first steps". After everything, it was easy to see what appeared from his power.

From what was created, and fashioned, it eventually appeared. A name was given to everything that was formed. Hence, he showed the differences among the other ones that were similar to himself, who have no father, from the beginning to the end.

They are all at rest in him and they are satisfied in Him. Their race or group is called "The Generation Over Whom There Is No Kingdom." They have glory without number. They unite themselves, by the use of his name.

They are your friends and your associates. They are all at rest in him, ever rejoicing in infinite joy in his unchanging glory and measureless jubilation. He is full of ever-imperishable glory and unexplainable joy. This was never heard of or known among all the angelic groups and beings and their worlds until now.

He made seventy-two angelic beings appear in the incorruptible generation, in accordance with the will of his own Spirit. They were uncompromised light or photon based life forms with redeemable souls.

The seventy-two luminaries or angelic beings, made three hundred and sixty other associates, angels, or luminaries, appear. They were an incorruptible generation, in accordance with the will of the Spirit.

Non Human, Light Energy Adamas, or First Light

Non Human, Light Energy Adamas, or First Light could have been made of half man, and one half light. Primordial Adam may have had a shape of a human being.

This configuration could have been the standard pattern for all of the life forms that followed after him. Adamas was assembled out of the first light, or luminous cloud that no angel has ever seen before.

Adamas simply means "human being". This light based, life form is not an actual human, but he is a light energy pattern, or Man of Light. The primordial Adamas is beyond our comprehension and he is made up of uncountable worlds of light.

The Primordial Adamas is mostly nonsexual or androgynous. He is neither male nor female, but both. His body and his mind are beyond our understanding.

His body is more creative and more versatile compared to a human body since his motions, and existence, exceeds 186,000 miles per second. His image is beyond the resolution of the human eye.

The light-energy Adamas is variable and multifaceted, since his body structure is somewhat human, somewhat angelic, and it is a powerful mechanism. Some of his internal parts are called souls.

Souls sometimes become corrupt because they forget their original birthplace. Human souls are similar to the primordial Adamas' soul. Each soul retains its likeness with Adamas. He is the typical pattern for all men that exist.

His eyes, ears, face, arms, hair, and legs may be thought of as to have a unique power of persuasion and existence. However, if he dwells in the human world, his behavior might become altered or compromised by evil beings.

The First Human That Evolved

It is widely thought of, believed, and proven, that eons before our civilization, that an evolution began as a result of random luck. Men evolved on the earth, just as the theory suggests. This fact is widely agreed upon and widely accepted by most.

It is also believed that the first man, who was born in the image of an angelic deity, also walked the earth. He is not to be confused with the evolutionary man and not to be confused with Adamas. <u>He was Adam.</u> According to an Italian farmer named Bart, Adam was 120 feet tall (eighty cubits) and his female associate, Eve, was 75 feet tall. (fifty cubits)

The primary one and all of his associates were still suffering from acute anxiety, loneliness and emptiness. The primary one decided to create a man in their own likeness, in order to increase their jubilance.

So, The Primary One sent Michael on an errand. He said to Michael; bring me lumps of soil from the four corners of the earth. At that time, those parcels of land were Ethiopia, Iraq, Algeria and India. He also said; bring me water from The Nile, The Euphrates River, The Tigris, and Gihon.

With all these ingredients, he formed the first man. The exact elements in the waters and in the soil that he had used to make a man are;

Elements
Carbon
Calcium
Phosphor
Potassium
Sulfur
Sodium
Chlorine
Magnesium
Oxygen
Hydrogen
Nitrogen

Sub Trace Elements
Arsenic
Iron
Iodine
Fluoride
Copper
Zinc
Chromium
Selenium
Manganese
Molybdenum

The First Body Was Born Dead

The body was motionless and did not move. It was simply clay mixed with water. Then, by means of transplant, a gift was given to the body. This gift was temporary. The gift could be easily taken back and was totally retractable and it was totally reconcilable. It was granted to the body on loan.

The gift was made of an indestructible, incorruptible, perpetual light formation, which would never perish. It was called a soul. This connection of the soul to the body, gave it life. It was at this time that the body was able to move and to breathe.

When the soul is removed, the body dies and returns to its original form which is made of clay and water. At any given time, the gift or the soul, could become detached and redeemable, regardless of the path the clay formation chose.

A secondary gift of free will was also granted to the lump of clay that was mixed with water. It was called spirit. It was at this tIme, in which the body of clay mixed with water, received a soul and it also received a self programmable spirit.

The body thrived and made extremely good use of its surroundings. The Primary One, His Son and all The Groups were very pleased.

LOOK ! What a beautiful life form. It has free will and it is partly formed of light. Based on its construction, it can live forever. The feeling of the dullness that existed in the world was finally overcome. The dullness was replaced with jubilation, happiness and joy.

They named him Adam
They named him in the honor of the first holy Adamas or First Light.

The Birth Of Hell

The entire star like groups, angels, stars or luminaries celebrated. They were jubilant over their new creation. However, the very first light based life form, who had the highest rank of the luminaries and was the greatest in beauty was nowhere to be found. He was made up of a handful of fire and he was called Satanael,

The Primary One asked Michael to summon Satanael so he could examine this new carbon based life form that brought happiness and jubilation to the luminaries of heaven. It was the will of the Primary One, that he and his highest ranking angel could celebrate this great birth of carbon together.

The conversation went something like this;

Michael said to Satanael;
Enjoy the new soil image of God, which he has made in the likeness of Himself.

Satanael said;
But I am fire of fire, I was the first angel formed, and you ask me to worship clay. Shall I worship clay and matter?

Michael said to Satanael,
Enjoy the clay made in Gods image, or he might be angry with you.

Satanael said,
God will not be angry with me? I will set my throne over and against his throne, and I will be as he is.

Michael then said to Satanael's 600 associates;
Enjoy the new soil image of God, which he has made in the likeness of Himself.

They said;
Just like as we have seen the first angel do, we will not worship him, which is less than ourselves.

<u>The Primary One</u> said to the 600 associates;
Enjoy the new soil image that I made in the likeness of myself, and be jubilant.

They said again;
Just like as we have seen the first angel do, neither will we worship him, that is less than ourselves.

The Primary One was angry with Satanael and his six hundred associates. He then commanded the windows of heaven to be opened. Then, the six hundred angels and Satanael were cast down by him. They were cast down upon the earth and they were senseless for forty years, and the sun shined seven times brighter than fire.

It was at this time that Satanael's name was changed from Satanael to Satanas. Suddenly Satanas woke up. He looked about and saw the six hundred others that were under him, senseless as well.

I (Satanael,) awakened my son Salpsan and asked his opinion on how I might deceive the man made of soil, on whose account I was cast out of the heavens.

As a result of his opinion, I took his advice and I did what he suggested. I took fig leaves in my hands and wiped the sweat from my bosom and below my armpits and cast them down beside the streams of waters.

I then took a vial in my hand and put them into the waters where the four rivers flowed out. Eve drank that water and desire came upon her. If she did not drink that water, I should have never been able to deceive her.

She Never Took A Bite
The Apple Was A Symbol

The evil powers were jealous of Adam and his spiritual partner because Adam spoke to Eve in a language that they did not understand. Their language was hidden from these evil powers. These adverse outsiders were deaf to Adam's communication. This gave Adam and Eve an opportunity to communicate in absolute and in unadulterated privacy.

The adverse authorities approached Adam. When they saw his female partner speaking with him, they became aroused and lusted after her. They said to each other; Come, let us ejaculate our semen in her, and they chased her. But, she laughed at them because of their foolishness and their blindness.

In their eyesight, she turned into an object that was as still as a tree, and when she left them, a shadow of herself that looked like her, remained in their mind. They defiled the shadow sexually, much in the same way a person may examine pornography and they temporally fulfilled themselves.

They defiled the sound of her voice. They did this in the same way a person may examine a pornographic soundtrack recording and temporally fulfilled themselves again. So, they enlightened and gratified themselves through the forms that they had shaped.

Eve separated from Adam because her friendship with him was not continuously obedient in the bridal chamber. Then, a female spiritual presence came to Adam from the spirit of the serpent; the devil. The devil tricked Adam and it possibly sexually seduced Adam. This may be the apple that Eve gave to Adam. (Separate and Private Lives).

There were two trees in the Garden of Eden. One bears animals or reptiles, the other bears men. Adam ate from the tree which bore animals, by having sex with them. He became an animal and he brought forth animals in the form of his own children. For this reason, one of his children worshiped animals.

Cain, who was born by Eve, was a murderer just like his father the serpent was. This was the same serpent who tricked Eve into having sex with him. Cain murdered his brother Able. Eve's action of fornication was not within obedience, and was enjoyed by her without emotion.

The partner of Adam was his soul. His soul eventually was taken away from him and so he died as a result of the soul being separated from the body.

Adam Is Removed From Hell
The following text is a rendition of the words that were found in a clay canister in Iraq. The text was deemed authentic by those who are educated and astute in carbon dating, and forensics. It was also authenticated by those who are authorities that study the slang language or the dialect of that language on the date in question.

This the testimony of Bart the farmer from Italy;

The Primary One said:
I went down into Hell that I might bring up Adam and all of them that were with him, according to the request of Michael the Archangel. Hell's associates said to the adverse one; we believe that a god comes here.

The angels spoke loudly to the chaotic powers and their associates, saying:
Remove your gates, you princes of darkness.
Remove the everlasting doors, for behold the King of Glory comes down.

Hell said: Who is this King of Glory that comes down from heaven?
The Primary One continued to explain. I had descended five hundred steps.
Hell was troubled.

As a group, they said;
I hear the breathing of the Most High, and I cannot endure it.
What is that suffocating, stinking, stench?
He comes here with a great a fragrance and I cannot bear it.

But the devil answered them and said;
Do not submit yourself to him.
O Hell, be strong, for God himself has not descended upon the earth.

The Primary One then said to Bart;
I had descended five hundred steps, and then the angels hollered out;
Take hold, remove the doors, for behold the King of glory comes down.

And Hell said: O, woe unto me, for I hear the breath of God.
And the Adverse One said to Hell:
It is Elias, or Enoch, or one of the prophets that this man seems to be.
But Hell answered Death and said:
Not even six thousand years have gone by yet.
What are they doing here?

The devil said unto Hell:
O Hell; the number of those years is in _my_ hands!
Hell, why are you afraid?

It is a prophet, and he has made himself appear to you to look like a God.
It is a prophet who mistakenly thinks he will ascend into heaven.
Hell said: Which of the prophets is it? Show me:
Is it Enoch the scribe of righteousness?

God will not make him suffer or arrive here until the six 6,000 years are up.
Do you think that it is Elias, the avenger?
But before the six thousand years are up, he is not supposed to come down.
What shall we do?
This destruction is of God.

Surely our end is at hand.
The devil said; I have the number of the years in my hands.
Don't be troubled, make safe and strengthen your bars and gates.
God does not come down upon the earth.

Hell then responded to the adverse one and said, what we hear from you is a lie. My belly hurts, and my inward parts are in pain. It can only be that God has come here. Where shall we hide before the face of the power of the great king?

Then, The Primary One caused havoc in Hell, breaking the doors, binding all of the demons in hell and he delivered Adam and all the holy souls upward. He redeemed the souls and then they ascended into Heaven. He then crowned Adam. After this ascension, there was a series of songs of welcome and a singing of happiness in heaven.

Adam was set at the gate of life to greet all the righteous souls as they entered heaven. Eve was set over all the women who had done the will of God, to greet them as they came into the city of The Primary One.

How many souls depart out of the world daily, Bart asked of The Primary One. Primary said to him, thirty thousand. Bart then said, Lord, if thirty thousand souls depart out of the world every day, how many souls out of them are found to be righteous? The Primary One said, hardly fifty.

Bart then said, Lord, how many extra souls are created daily? Primary said to him, one soul only is born above the number of them that die. (Equation... If 100 souls depart, 101 souls are generated. This equation suggests that the world population grows daily.) Bart said: How many souls depart out of the body every day? Primary said, twelve thousand, eight hundred and eighty three souls depart out of the body every day. 12,883.

Bart then asked The Primary One, if he could see the likeness of The One Who Is Most Adverse. Michael sounded a trumpet, and the earth shook. Satanael came up from hell, being held by 6,060 angels and he was bound into place with fiery chains.

His length was 2,300 feet, his width was 60 feet, his face was like the lighting of a fire and his eyes were full of darkness, like sparks. Out of his nostrils came a stinking smoke. His mouth was like the opening of a cave made of stone. One of his wings was 120 feet long.

Bart asked The Primary One, if He could question the devil, The One Who is most adverse. Then, The Primary One answered him and said;
You bold heart! You ask for that which you are not even able to look upon.

The Primary One said; Bart, you will see the enemy of men? Go near him, and trample on his neck with your feet and he will tell you his work, and how he deceived men. Go and place your foot upon the neck of Satanael. Bart put his foot upon his neck and then, Satanael trembled.

Bart said to him: Tell me who you are and what is your name. Satanas said to him: Lighten me a little on my neck, and I will tell you who I am, how I came here, what my work is, and what my power is. And Bart lightened his neck and said to him: Say all that you have done and all that you do.

And the devil answered and said: If you will know my name, at first I was called Satanael, which is interpreted as a messenger of God. When I rejected the image of God that was made out of clay and water, my name was changed to Satanas. That name means an angel that keeps hell. And again Bart said to him: Reveal to me all things and hide nothing from me.

And Satanas said to him: I swear to you by the power of the glory of God, that even if I would hide all, I cannot hide anything, for he that is near would convict me. For if I were able, I would have destroyed you, just like them that were here before you. For, indeed, I was formed and I was the first angel.

When God made the heavens, he took a handful of fire and formed me first. I would not worship Adam who was made of clay and water. The other archangels struck me with their rods and pursued me seven times in the night and seven times in the day.

These are the angels of vengeance which stand before the throne of God: These are the angels that were formed first, and after them were formed all the other angels.

Bart said to him: Do you pour chastisement upon the souls of men?
The devil said to him: You will be deceived in the punishment that I pour on them that are of the hypocrites, of the back-biters, of the jesters, of the idolaters, of the covetous, of the adulterers, of the wizards, of the diviners, and of them that believe in us, and of all whom I look upon.

Those souls that do these things, and consent to do these things of their own will, do perish along with me. Bart said to the devil: Declare quickly how you persuade men not to follow God. Declare it in few words.

The devil said: I will tell you. A wheel came up out of hell, having seven fiery swords. The devil ground his teeth together loudly. It was after this grinding noise when a wheel having swords that flashed with fire, and pipes came up out of the bottomless pit.

Bart asked him, saying: What is this sword?
The devil said:
1) This sword is the sword of the gluttonous or the greedy. It is into this pipe they are pushed. Through their gluttony they devise all types of sin.

2) Into the second pipe of fire are sent, those who make spiteful or slanderous comments about somebody who is not present.

3) Into the third pipe of fire are sent the hypocrites and the rest whom I overthrow by my plots and schemes.

4) Into this pipe of fire are put the fortune tellers, the card readers, the casters of lots, all the diviners, the enchanters, and all of those who believe in them, or all of those who have wandered around looking for them. Those who believed in them are cast there because they have invented these false divinations by looking around for them in the first place.

5) Into this pipe of fire are first the blasphemers ... suicides ... idolaters....

6) And (7) in the rest of the pipes are those who swear to tell the truth under oath and then lie. Besides these, there are large amounts of the guilty.

Bart said: Do you then do these things by yourself alone?
And Satanas said: If I were able to go forth by myself, I would have destroyed the whole world in three days, but neither I nor any of the six hundred actually go forth. For we have other swift ministers that do what we want them to do and whom we command.

We furnish these ministers of ours with a hook that has many points and send them forward to find these sinners. They catch for us the souls of men, enticing them with the sweetness of baits that a diver would use such as, drunkenness, laughter, by backbiting, hypocrisy, pleasures, fornication, and the rest of the vices.

The devil ambiguously said: I will tell you also the rest of the names of the angels that are God's associates, that are in heaven and they do help us do this as well. They are traitors and guilty of treason.

Bart said: Be still (be muzzled) you dragon of the pit.
Again the devil ambiguously said: Many things I will tell you about the angels that run together throughout the heavenly places and the earthly places and who they are.

Bart said to him: Be still, be muzzled and be faint. Bart released Satanas and said to him: Go to your place, with your angels, but the Lord has mercy upon his entire world.

Satanas complains enormously amplified in the language of Latin that he has been tricked into telling his secrets before their due time.

Bart had never seen anything to compare with the beauty and the glory of the Adam who was the one saved from hell by The Primary One. Adam was forgiven, and all the angels and saints rejoiced and saluted him. After this, they then departed, each to their own place.

In Bart's testimony, he desires to be humble. He protested: I am the least of all of you. I am a humble workman. When they report my testimony, I want them to say, "How can this be. Isn't he Bart the gardener and the dealer in vegetables who was born in Italy?"

Who Is The Fake Begetter?
This Portion Of The Documentary Is The Approximate Writings Of The Finest Minds Regarding Indestructible, Photonic, Light Based, Life Forms.

Imagine an interplanetary, cosmic deity made up of indestructible particles of light. He thinks of himself as being the only god in the entire cosmos. He suffers from cosmic egotism. His thoughts are false wisdom, and lies.
He is referred to as Fake God or Fake Begetter or Yaldabaoth.

He cannot understand the galaxies nor can he understand true wisdom. He cannot understand the cosmic world that came about from the explosion of the unnamable substance that produced him in the first place.

He becomes puffed up, and bloated with the illusion, that he is truly wise. Angelic deities feel shame and want to hide him from themselves. Wisdom shuns him away from her radiance, so that no one who is immortal might encounter him.

Wisdom is a light orchestrated, star like, incorruptible, and indestructible sequence of beings with its thrones in the middle of the cosmos. The Cosmic or Angelic Group, better know as wisdom, is that cosmic force that organizes the cosmos and produces the true life and true understanding.

Wisdom is the galactic core of the cosmos, but it does not cause the birth of the core itself. The process of these galactic compressions and galactic explosions takes place continually causing galactic arms or star like branches.

This intergalactic process is constantly manufacturing light and matter, producing new stars, new worlds and new planets.

Yaldabaoth and his associates, have no creative capacity of their own. They can certainly mimic or imitate whatever they see. They imitate what they see so well, that they have convinced themselves with the allusion that they can create life itself.

Yaldabaoth is so disillusioned, that he sees himself as the top solar deity. This counterfeit cosmos that he produces is sizeable when one thinks of it on a planetary basis. He actually creates nothing. He is the architect of lies. What he makes is quite sterile and empty as what he makes lacks soul.

He attempts to make a parallel universe in which there are celestial bodies, heavens and angels. He actually constructs children, wives and wars to be won for the shell-like world he claims that he has mastered. He calls himself, the god of gods.

The galactic community sees him as a nuisance and that he threatens the growth and he threatens the prosperity of the cosmos. He is referred to as Fake God or Fake Begetter

THE GOOD

During the course of history, many were distinguished by humility, and were in no respect puffed up with pride. They yielded to obedience rather than ignored it. They were more willing to give it, rather than to receive it. They were content with the provisions that were made available to them.

Many were inwardly filled with good doctrine thus; a profound abundance of peace was given to them. They had an unquenchable desire for doing good. Day and night they were anxious for the good of their neighbors, and that the number of the good, might be saved with the use of mercy and good conscience.

They were sincere, uncorrupted, and forgave the injuries between one another. Fiction was eradicated. They mourned over the transgressions of their neighbors. They took fault for their own deficiencies and they deemed their faults as their own.

They never held a grudge. They were always ready to do good work. Dressed with a thoroughly honorable and a religious life, they did all these things in the fear of God. The commandments and the ordinances of the Primary One were written upon the tablets of their hearts. Every kind of honor and happiness was given to them.

How to live

Let us say then, from the heart, that you are reasonable. Speak of the truth, with those who search for it. Make it known to those who have committed their sin in error. Make firm, the foot of those who have stumbled, and stretch out your hands to those who are ill.

Feed those who are hungry, and give comfort to those who are weary. Rise up those who wish to rise, and awaken those who wish to sleep. Do not strengthen those who are obstacles to you, or to those who are collapsing on you, as though you were a support for them to lean on.

With sufficient clearness, The Primary One distinguishes the three classes of beings. They were the elect, the lost, and those who were remaining in faith. They shall have confidence and not be confounded. Confusion is a heavy punishment for those who are in error.

Is Death Real

The Words Of Ms Magdalene of Magdala
These are the approximate words of Ms Magdalene of Magdala. She kept her records on papyrus paper, which was scientifically carbon dated and written in the year 3 A.D. This was several thousand years ago.

The language that she used was also verified as being authentic by scholars who study the phonetic dialect and the slang that was used during that time. These are some of the things The Primary One told her.

The Soul And The Seven Powers;
The Primary One said to her;

All natures, formations, and creatures exist in each other and with one another. They will untravel, and again return into their own roots. For the nature of matter, is to be resolved and to return back into the roots of the nature alone that it came from.

There were seven angry powers of wrath which took seven forms. Their formation was darkness, desire, ignorance, the fear of death, the kingdom of the flesh, the foolish wisdom of the flesh, and the wrathful wisdom. These are the seven powers of wrath.

The soul from an expired body came up to the first authority.
The first authority was called darkness. Power questioned the soul. The soul was addressed and accused something like this. Oh soul....Why do you dare to ascend? It is darkness you come from, it is the darkness that you caused and now you are attempting to spread and carry this darkness forward.

The soul said, what you have said may be true but, the darkness wasn't totally dark. There was brightness mixed in with the darkness and I shall move forward and carry the brightness forward with me and leave the darkness behind me.

The soul from an expired body came to the second angry authority. It was called desire. Soul was addressed like this. It is desirable that I did not see you descending, but now I see you ascending. Then the angry power said, Oh! Soul, why do you lie, since you belong to me?

The soul answered and said, I saw you, but you did not see me nor did you recognize me. I desired to serve you as a garment, and you did not notice me. When the soul had said this, it went away rejoicing greatly.

Again the soul came to the third angry authority, which is called ignorance. Power questioned the soul, 'Where are you going? You who are bound to be wicked and not bound to judge others!

And the soul said, why do you judge me although I have not judged anyone? I was bound even though I did not bind others. I was not recognized. But I have recognized all that is being dissolved, both the earthly things before now and the heavenly things.

The angry powers asked the soul, where do you come from, "You" you that are a slayer of men, and where are you going, you conqueror of space?

The soul answered and said,
What binds me has died. What surrounds me, I have overcome, and my evil desires have ended. My ignorance has died! I was released from a world, and brought into a new form of a heavenly type and of a new heavenly shape that is different from the one I once had.

I was released from the chains of stupidity which is a brief life. Starting now, I get the rest of time, of season, and of obviousness, to be in silence. The soul pleads in a silent way. It is now less corrupt, more humble and it has accepted the idea that it shall be patient and await the decision of its destiny.

These are the approximate words of Ms Magdalene of Magdala and what The Primary One said to her.

Who & What Is The Primary One

The Primary One is as old as the light around him, but the light around him has not acquired or exceeded his knowledge. He is immortal, eternal, and has no birth that a man's conscience can grasp. He has no known father and he has no known beginning that man can understand.

He sees himself as a First Existent, and an Un-Begotten Father. He is equal to the age of the light that is before him, but the light is not equal to him in power. What is his name? He has no recollection of the past since the past is a distance which is beyond the scope of thinking.

At the first formations of the crushing weight of unknown substance and explosions, causing the first light, He found that he was alone. There was no need for a name since no one spoke to him. He did not receive a name as others do. He is indescribable and unnamable, until a time finally came when one who was his son and was also an entity, spoke to him.

It his son, who has the power to speak his name. It pleased him that his name, which is loved by his son, should be known to his son. He gave the name to his son. He spoke to his son about his secret things. His son knew that the Father is a being, without evil.

For that very reason, he brought his son forward, in order to speak about his resting place. It was the son who had come forward, to glorify the divine powers, the greatness of his name, and the sweetness of <u>His Father.</u>

His own resting-place is an unidentifiable place or heaven. Therefore, all the thoughts of the Primary One are absolute, and they are divine thoughts. His associates possess his head, which is a place of rest for them. They are supported while approaching him, and they have participated in his face by means of kisses.

But they do not become totally clarified in this way, for they themselves are not exalted. They did not despise The Primary One, nor did they think of him as small, nor that he is harsh, nor that he is wrathful, but rather that they thought of him as being without evil, imperturbable, sweet, knowing all spaces before they have come into existence, and he had no need to be instructed.

I want you to know that all men that are born on earth, and from the foundation of the world until now, are dust.

Man has inquired often about God, regarding whom he is, what he is like, and they still have not found him. Now the wisest among them may have speculated about the order of the world and its movement.

But their speculation has not reached the truth. It is said by all of these philosophers that the order of the world is directed in three ways. As a result, they all disagree with each other. Some of them say it is directed by itself. Others say that it is directed by providence. Others say that it is directed by destiny.

But it is none of these. Of the three voices I have just mentioned, none of them is close to the truth, and these thoughts are from man. Whatever is from itself is a polluted life; it is self-made. Luck has no wisdom in it at all. Chance does not discern. Knowledge of reality is given to whoever is worthy to receive it.

Some of the ones I speak of have not been fully defined. They are an undying chaotic entity, who exists in the same presence and the same area where mortal men dwell, or in the mist of men that perish.

He, who is beyond words, has principles that can't be measured. No authority is over him. He is subjected to no one. There is no creature with his understanding starting from the foundation of the world until today, except for himself. Anyone who he wants to have new understandings, gets knowledge from First Light.

He looks to every side and sees itself from himself. Since He is infinite, he is incomprehensible and he is not understandable. He is imperishable. He will never die. He has no likeness to anything. He is the unchanging good!

He is faultless. He is eternal and lasts forever. He is blessed. While he is not known, he ever knows himself. He is immeasurable. He is not traceable. He is perfect, having no defect. His blessed imperishability, will never end. He is called   """Father of the Universe""".

All his powers are equal. Before anything is seen, of those who can be seen, majesty and authority are in him. He embraces the whole of all of the totalities, while nothing embraces him. For he is all mind. He is thoughtful, considerate, reflective and a rational power. All of these are equal in power, in him.

His power is the source of all of the totalities. Whole races, from first to the last were in his knowledge and in his infinite understanding. He knew about them before they began. He is an unbegotten Father who has no father. This great wealth that was hidden in him, might be revealed to you.

Because of his mercy and his love, he wished to bring forward fruit by himself. However, He does not deliver his goodness alone, but other spirits of the Unwavering, Incorruptible, Generational, type, may aid him.

They might bring forward body, fruit, glory, and honor that is indestructible and respects his infinite grace. His treasure might be revealed by himself. He is a Self-begotten God. He is a God with no father. He is the father of every life form that will not die and that came into existence afterward. But they had not yet come to visibility.

Now there is a great difference that exists among the imperishable, also known as those that will not die. Everything that was made by the perishable will perish, since it came from the perishable. But whatever came from imperishableness, uncorrupt, undying ones does not perish. It also becomes imperishable.

So, many men went astray because they had not known this difference and they died in the body and in the spirit. Look into what is visible to you, in order to find what seems to be invisible to you. If you just think, you will find comfort in some of the things that are invisible to you. They can actually be found in those things that you can see. These invisible things belong to the Un-begotten Father.

The Lord of the Universe is not called 'Father', but 'Forefather'. He is the beginning of those that ever appeared later. Seeing himself in a mirror, he appeared to be a resemblance of himself. His likeness appears as A Divine Self-Father, Confronter over those who are confrontational.

I want you to know that He, who is A Self-constructed Father, came to life or appeared before the universe and its infancy. He is Self-grown, full of shining light and ineffability. At the very beginning, he decided to have his likeness become a great power. That Principle Light appeared as an Immortal, never dying, genderless Man.

First Man is called 'Begetter, Self-perfected Mind'. He pondered with The Great Philosophy and The Great Wisdom, who is his companion, and revealed his thoughts to his first-begotten, genderless son. His male name that was assigned to him was, 'First Begetter, Son of God', his female name that was assigned, was 'First Be-Gettress, Sophia, Philosopher, Mother of the Universe'.

He created a multitude of angels, or followers without number from the Spirit and from the light that was around him. The First Man, Father, or Begetter who has a father is called Adam, which means "Eye of Light," because he came from a shining Light.

His holy entities, are unexplainable, shadow-less, and will forever rejoice with joy, in the brilliance which they received from their Father. The whole Kingdom of Son of Man, who is called Son of God, is full of indescribable and shadow-less joy, and unchanging jubilation.

Your roots are in the infinities that do not die. When those whom I have discussed earlier were exposed, the Self-Begetter Father, very soon created twelve groups, for the twelve angels. All these groups are perfect and good.

However; regarding these rebellious, adverse, authorities, a defect in the female appeared. One group, is that of Man, in which there is no kingdom, who is called 'Adam, Eye of Light'. The second group embraces the Eternal Infinite God, who is the Self-begotten group and the groups that are of him.

Some exposed the groups of chaos, and the kingdoms of chaos. They wanted their authority to appear in man, that they might exercise their desires until the last things that are above in the heavens, are chaos. It was their will to use man as a tool in order to corrupt the elite ones.

The powers of chaos very soon appeared and were called gods. Their wisdom revealed lords; and the lords from their power revealed archangels; the archangels from their words revealed angels; from them, semblances appeared, with structure and form and a name for all the groups and their worlds.

They created hosts of chaotic angels, without number. They created new or virgin spirits that are unexplainable and unchangeable lights. They have no sickness, nor weakness.

The area from each group or world that came from it, took a pattern for their own creation. This is the likenesses in the heavens of chaos and in their worlds. All natures, starting from the revelation of chaos, are in the Light that shines without a shadow.

They seem to also have joy that cannot be described, and unutterable jubilation for themselves. They ever delight themselves and congratulate themselves on account of their unchanging glory and their immeasurable rest.

The Mother of the Universe and her companions, desires by herself and without males, to bring those of chaos into existence without these male companions. The will of the Father of the Universe, who opposes her, is that his unimaginable goodness might be exposed through her badness.

He created that curtain between the immortals and those that came afterward. A consequence might follow ... every group and every chaos and every defect of the female might come about that Error, and make her content. These became the blind spirits. From the groups above came an emanation Light.

Bent-Begetter is called 'Yaldabaoth' or fake god.
These names were received by all who are in the world of chaos.

The names are.

Doom,

Old Age,

Retribution,

Death,

Sleep,

Strife.

Hell...Q & A

Interviews With The Seers

(Excerpts taken from The Gaspa at Medjugorje)

Interview With Mirjana Dragicevic

M: I have seen The Gaspa for eighteen months now, and feel I know her very well. I feel she loves me with her motherly love, and so I have been able to ask her about anything I would like to know. I have asked her to explain some things about; Heaven, Hell and Purgatory.

Hell was not clear to me. For example, I asked her how God can be so unmerciful as to throw people into Hell, to suffer forever. I thought: If a person commits a crime and goes to jail, he stays there for a while and then that person is forgiven, but to Hell, forever? Why?

She told me that souls who go to Hell have ceased thinking favorably of God and have cursed him, more and more. So they have already become a part of Hell, and choose not to be delivered from it.

Then, She told me that there are several levels in Purgatory: Some levels are very close to Hell. These levels extend higher and higher and migrate closer and closer toward Heaven. She said, that most people think that many souls are released from Purgatory and go to Heaven on All Saints' Day, but most of the souls are taken into Heaven on Christmas Day.

T: Did you ask why God allows Hell?

M: No, I did not. But afterward I had a discussion with my aunt, who told me how merciful God is. So I said, I would ask the Madonna about this question.

T: According to what you have said then; it's as simple as this: People who oppose God on earth just continue their existence after death, and oppose God in Hell afterward?

M: Really, I thought if a person goes to Hell...Don't people pray for their salvation? Could God be so unmerciful as not to hear their prayers? Then the Madonna explained it to me.

People in Hell do not pray at all, instead, they blame God for everything: In effect, they become one with that Hell. They rage against God, and they suffer, but they always refuse to pray to God.

T: To ask Him for salvation?
M: In Hell, they hate Him even more.

T: As for Purgatory, you say that souls who pray frequently are sometimes allowed to communicate, at least by messages, with people on earth, and that they receive the benefits of prayers said on earth?

M: Yes. Prayers that are said on earth for souls who have not prayed for their salvation are applied to souls in Purgatory.

T: Did the Madonna tell you whether many people go to Hell today?

M: I asked her about that recently, and She said that, today, most people go to Purgatory, the next greatest number go to Hell, and only a few go directly to Heaven.

T: Only a few go to Heaven?

M: Yes. Only a few, the least number go to Heaven.

T: Did you ask about the conditions for a person to enter Heaven?

M: No, I didn't; but, we can probably say what they are, that God is not looking for great believers but simply for those who respect their faith and live peacefully without malice, meanness, and falsehood.

T: This is your interpretation, your understanding?

M: Yes. After I talked to the Madonna, I came to that conclusion. No one has to perform miracles or do great penance. They merely have to live a simple, peaceful life.

T: Well, besides Heaven, Hell, and Purgatory, is there anything else that is new recently?

M: The Madonna told me that I should tell the people that many in our time judge their faith by their priests. If a priest is not holy, they conclude that there is no God. She said, you do not go to church to judge the priest, or to examine his personal life.

You go to church to pray and to hear the Word of God from the priest. This must be explained to the people, because many turn away from their faith because of priest's inconsistencies.

Today, as it was a long time ago, the Virgin told me that God and the devil conversed, and the devil said that people believe in God only when life is good for them. When things turn bad, they cease to believe in God.

Then, people blame God, or act as if he does not exist. God, therefore, allowed the devil one century in which to exercise an extended power over the world, and the devil chose the twentieth century.

Today, as we can see all around us, everyone is dissatisfied. They cannot abide each other. Examples are the number of divorces and abortions. All this, the Madonna said, is the work of the devil.

T: This behavior of people, they are under the influence of the devil, but the devil does not have to be in them?

M: No, no. The devil is not in them, but they are under the influence of the devil, although he enters into some of them. To prevent this, at least to some extent, the Madonna said we need communal prayer, and family prayer.

She stressed that there is the need for family prayer most of all. Also, every family should have at least one sacred object in the house and the house should be blessed regularly.

She also emphasized the failings of religious people, especially in small villages, for example, here in Medjugorje, where there is separation from Serbians and the Muslims.

This separation is not good. The Madonna always stresses that there is but one God, and that people have enforced an unnatural separation.

T: What, then, is the role of Jesus Christ, if the Muslim religion is a good religion?

M: We did not discuss that. She merely explained, and deplored the lack of religious unity, "especially in the villages." She said that everybody's religion should be respected, and, of course, one's own religion should be respected.

T: Tell me where the devil is especially active today. Did She tell you anything about this? Through whom or what does he manifest himself in the most?

M: Most of all through people of weak character, who are divided within themselves. Such people are everywhere, and they are the easiest for the devil to enter.

But he also enters the lives of strong believers, sisters, for example. He would rather corrupt real believers than nonbelievers. How can I explain this? You saw what happened to me. He tries to bring as many believers as possible to himself.

New Interview

S: When you talk with the Madonna in your visions, we cannot hear you speak.

Ivanka: We speak out loud, the same as now.

S: Let me put it this way. Do you speak with the Madonna mentally—that is She understands what you think, or do you speak to her in a low voice, a whisper, so we cannot hear you? Or is your conversation miraculous, beyond our power to hear and understand?

I: I speak with her normally, the same as I'm speaking now. Also, I hear her voice and her words in the normal way. I also hear as well, what the others in the group say to her.

S: Did you ever ask for anybody in your family who has passed away?

I: Yes, I asked not long ago, and she said the same thing.

S: What did Our Lady say?

I: She said my mother is with her, and that I should obey and not worry. We asked her why, of all the people that are around, why did she appear to us. She said; She does not always seek out the best people.

S: To whom are the Madonna's messages sent?

1: To the whole world.

S: What are the messages?

I: Peace, conversion, fasting, penance, and prayer.

S: Which is the most important?

I: Peace.

S: Why peace?

I: When everyone in the world is at peace, everything is possible.

S: Bearing in mind what you know about the future, tell me if the Madonna of Medjugorje will reconcile the world even more.

I: I think She will.

S: You speak to the people of the Madonna's messages, but some do not believe you. In that case, what do you do?

I: I pray for them, that God will enlighten them.

S: Can you do anything else?

I: I will go on trying to persuade them. They will be convinced.

S: There are those who are opposed to the Madonna. What would you tell them?

I: I'd tell them: Convert! There is a God. That is it!

S: Can those who oppose the Madonna, frustrate her plans in the world?

I: No, All their armaments and explosives could not destroy her plans.

S: Can those who oppose the Madonna do harm to the souls of the people and to the Madonna's plans for the souls of the people?

I: No.

S: Does that mean the Madonna is stronger than they are?

I: Normally She is. It is Jesus who decides—God. Not the Madonna.

S: That means, She cannot act independently on her own. She must do the will of God, as she always has.

I: I believe so.

S: It is important that people of good faith, regardless of their denomination, should not be turned against each other. But, tell me more about this. What did the Madonna say about this?

I: The Madonna said that religions are separated on the earth, but the people of all religions are accepted by her Son.

S: Does that mean that all people go to Heaven?

I: It depends on what they deserve.

S: Yes, but many have never heard of Jesus. Can they go to heaven?

I: Jesus knows all about that; I don't. The Madonna said, basically, religions are similar; but many people have separated themselves from each other because of religion, and they become enemies of each other.

This Interview Ends

The visionaries usually describe Our Lady as wearing a long gray dress, with a white veil. She has blue eyes, long black hair, and rosy cheeks.

Our Lady's dress is of a certain color that does not exist on this earth. It has something to do with the color gray, so we say: 'Her dress is gray.' This observation that I made is the same for all of the rest of us.

We have some words that we use to describe the Madonna, but everything in the experience of Our Lady is beyond words.

How do they describe Purgatory?
Marija says, that Purgatory is a large place. It is foggy. It is ash gray. It is misty. She said that you cannot see people there. It is as if they are immersed in deep clouds and they are immersed in a burden of suffering.

The biggest suffering that souls in Purgatory have is, they see there is a God, but many times they did not accept Him here on earth. Now that they are in purgatory, they long so much to come close to God. Now they suffer intensely, because they recognize how much they have hurt God, how many chances they had on earth, and how many times they disregarded God.

Vicka said that she couldn't see people in Purgatory, but she could sense people weeping, moaning, and trembling in what seemed to be like a terrible suffering. Yes, Purgatory exists with a suffering in a way that can never be expressed in a thousand lifetimes on earth.

Imagine the terror of being strangled to death for a minute or two, and then, it is over. But imagine an elongated strangling for months or even years. Vicka used the word "strangling" as one aspect of the suffering in Purgatory. This is a ghastly suffering, and this knowledge should give us an impulse to pray for the poor souls that are in Purgatory.

A special note-worthy to know about Purgatory is, when you pray for a loved one by name who is in Purgatory, they can see you on earth, at that moment.

Jakov said, "Hell is the ultimate waste because no one needs to go there." Vicka describes hell in the following manner: In the center of this place is a great fire, like an ocean of raging flames. We could see people before they went into the fire, and then we could see them coming out of the fire. Before they go into the fire, they look like normal people.

The more they are against God's Will; the deeper they enter into the fire. The deeper they go into the fire, the more they rage against Him. When they come out of the fire, they do not have a human shape anymore. They are more like grotesque animals, but unlike anything on earth.

When they came out of this fire, they were raging, smashing everything around, hissing, grinding their teeth and screeching. And yes, hell exists. If you saw hell as Jakov and Vicka have seen hell, you would not want your worst enemy to go there.

Today, I was led by an Angel to the chasms of hell. It is a place of great torture. How awesomely large and extensive it is! These are the kinds of tortures I saw: The first torture that constitutes hell is the loss of God; the second is a perpetual remorse of conscience;

The third torture is that one's condition while in hell will never change. The fourth is a fire that will penetrate the soul without destroying it. It is a purely spiritual fire, lit by God's anger. The fifth torture is a continual darkness and a terrible suffocating smell. Despite the darkness, the devils and the souls of the damned can see each other. They can see the evil of their own soul and they can see the evil of the others.

The sixth torture is the constant company of Satan. The seventh torture is a horrible despair, a hatred of God, vile words, curses and blasphemies. These are the tortures suffered by all the damned together, but that is not the end of the sufferings. There are special tortures destined for particular souls.

These tortures are the torments of the senses. Each soul undergoes terrible and indescribable sufferings, related to the manner in which they have sinned. There are caverns and pits of torture where one form of agony differs from another. I would have died at the very sight of these tortures if it wasn't for the power of God supporting me.

Let the sinner know that the senses in which he made use of to sin, that he will be tortured with those senses throughout all eternity. I am writing this at the command of God, so that no soul may find an excuse by saying there is no hell, or that nobody has ever been there, and I wrote this so no one can say they didn't know what it is like.

Vicka: States;
We saw many people in hell. Many are there already, and many more will go there when they die. The Blessed Mother says that those people who are in hell are there because they chose to go there.

They wanted to go to hell. We all know that there are people on this earth who simply don't admit that God exists, even though He helps them. He gives them life, sun, rain and food. He always tries to nudge them onto the path of holiness.

They just say they don't believe, and they deny Him. They deny Him, even when it is time to die. And they continue to deny Him, after they are dead. It is their choice. It is their will that they go to hell. They choose hell.
People turn away from God by the choices they make.

In this way they choose to enter the fire of that hell, where they burn away all their connections to God. That is why they can never get back to God. It takes God's mercy to get back to Him. In hell, they no longer have access to God's mercy. They choose to destroy their beauty and their goodness. They choose to be ugly and to be horrible.

People do this all the time. Each choice that they make is against God, against God's Commandments, and against God's Will. They become one with hell even while they have their body and while they are alive. At death, they go on as they were when they had a body.

Marija has shared this about hell:

Question: "Marija, have you ever seen hell?

Marija: "Yes, it's a large space with a big sea of fire in the middle. There are many people there. I particularly noticed a beautiful young girl. But when she came near the fire, she was no longer beautiful. She came out of the fire like an animal. She was no longer human.

The Blessed Mother told me that God gives us all choices. Everyone responds to these choices. Everyone can choose if he wants to go to hell or not. Anyone who goes to hell, has chosen hell.

Question: "Marija, how and why does a soul choose hell for himself for all eternity?

Marija: "In the moment of death, God gives us the light to see ourselves as we really are. God gives a freedom of choice to everybody during his lifetime on earth. The one who lives in sin on earth, can see what he has done and recognizes himself as he really is.

When he sees himself and his life, the only possible place for him is hell. He chooses hell, because that is what he is. That is where he fits. It is his wish to go to hell. God does not make the choice. God condemns no one. We condemn ourselves. Every individual has a free choice. God gave us freedom.

Question: Marija, what about people who grow up who are spiritually deceived, people who have been told that God does not exist, that there is no God?

Marija: People, as they grow up, can think. Everyone knows and can recognize what is good and what is bad by the time they grow up. God gives us freedom of choice. We can choose good or bad. Everybody chooses here in this lifetime, whether he goes to Heaven or hell.

Question: "**How do we choose Heaven or hell** or Purgatory for ourselves?

Marija: "At the moment of death, God gives everyone the grace to see his whole life, to see what he has done, to recognize the results of his choices on earth.

And each person, when he sees himself in the divine light of reality, chooses for himself where he belongs. Every individual chooses for himself what he personally deserves for all eternity.

Vicka
Jesus had long hair, brown eyes, and a beard. We only saw His head. Jesus appears to them crowned with thorns, and with injuries all over. The children are afraid.

Question;
Theologians have said, these would be the last apparitions on earth? Is it true?
The visionaries:
"These apparitions are the last in the world."

We go to Heaven in full conscience, that same conscience which we have now. At the moment of death, we are conscious of the separation of the body and of the soul. It is false to teach people that we are reborn many times and that we pass onto different bodies.

One is born only once. The body which is drawn from the earth, decomposes after death. It never comes back to life again. Man receives a transfigured body. Whoever has done very much evil during his life can go straight to Heaven if he confesses his inconsistencies, if he is sorry for what he has done, and if he shares food with another that is eaten in the memory of Jesus at the end of his life.

Today, many people go to Hell. God permits his children to suffer in Hell due to the fact that they have committed grave and unpardonable sins. Those who are in Hell, no longer have a chance to know a better life.

Men who go to Hell no longer want to receive any benefit from God. They do not repent nor do they cease to swear and to blaspheme God. They make up their mind to live in Hell and do not at all contemplate leaving it. Man's refusal is an irreversible choice.

Questions answered on the subject of Purgatory

There are different levels of which the lowest are close to Hell and the highest levels of purgatory gradually draw near to Heaven. The greatest numbers of souls that leave Purgatory is not on All Souls Day, but at Christmas time.

In Purgatory, there are souls who pray passionately to God but, no relative or friend prays for them on earth. God makes them benefit from the prayers of other people. It happens that God permits them to manifest themselves in different ways, close to their relatives on earth.

Things that fall, are things that can easily be stood upright again. The Primary One will come to a deficient person, and he shall bring the deficient person back to Himself once again.

For the bringing back is called 'repentance'. An agenda called Incorruptibility, breaths forward and pursues the one who sinned. It does this in order that the sinner might rest. Forgiveness is what remains in the deficient one, and in the word of the divine powers.

Peter of Galilee Complains About Hell
Peter complained to The Primary One and said;
Lord, sometimes you urge us on to the Kingdom of Heaven with promises, and other times you turn us away with threats. Sometimes you persuade us and drive us to faith and promise us life, and other times you expel us from the Kingdom of Heaven. So which one is it ?

The Primary one answered and said;
I have revealed myself to you. I see you rejoicing many times, and when you are thrilled with the promise of life, you are still gloomy. You are distressed when you are taught about the Kingdom.

Therefore, SCORN rejection when you hear it, but, when you hear the promise, BE THE MORE GLAD. In truth, I say to you, the one who will receive life and who believes in the Kingdom, will never leave it not even if My Father desires to banish him! Blessed is the one who has seen himself as a quadrant or a portion of Heaven.

What Is Compared To The Soul?
It is a precious and un-contemptible thing and it came to be in a contemptible body. Some are afraid that they might rise naked. Because of this, they wish to rise in the flesh. They do not know that it is those who wear the flesh, are the ones who are naked.

The superiority of man is not obvious to the eye. He has mastery over the animals which are stronger than he is. This allows him to survive. Men ate one another because they did not find any food. But now they have found food, because man has learned to till the soil.

The forms of an evil spirit include male ones and female ones. The males will unite with the souls of a female form. The females will mix with a male form, even though the males are not interesting to them. Males cannot escape females. The female will detain him.

When wanton women can see a male sitting alone, they leap down on him and play with him and defile him. Lecherous men do the same thing, when they see a beautiful woman sitting alone. They persuade her and compel her, wishing to defile her.

He who comes out of the world, is no longer detained by the grounds that were in that world, and is above the desire of complications and fear. He is a master over these things. He is superior to envy. If he gives into envy, it will seize him and throttle him downward. How will he be able to escape these great jealousies?

How will he be able to avoid them? There are some who say, that we are faithful, in order for us to escape the unclean spirits and the demons. If one has the Holy Spirit in them, no unclean spirit would cling to them. Fear not the flesh nor love it. If you fear it, the flesh will gain mastery over you. If you love the flesh, it will swallow you up and it will paralyze you.

There is an evil world after this world, which is truly evil. Death is called "the middle". While we are in this world, it is fitting for us to acquire the resurrection. We will strip off the flesh we are wearing. We may be found in death. For many go astray on the way. You can exercise your will in either direction.

What is Soul and what is the Spirit

It is not the flesh which yearns for the soul. For without the soul, the body does not sin. The soul is not saved without the Spirit. But if the soul is saved, the spirit also is saved. The body becomes sinless. It is the spirit which gives life to the soul. It is the body which kills it.

Do you imagine that many have found the Kingdom of Heaven? Blessed is the one who has seen himself as a quadrant or a portion of Heaven. Why I say this to you is so that you may examine yourselves. The Kingdom of Heaven is like a stem which has sprouted up in a field. And when it is ripened, it scatters its fruit and, in turn, it fills the field with more stems and plants for another year.

Work hard to simulate this style of plant, in order that you may be filled with the Kingdom. Do this so that the Black One may find no means of an entrance into your life. Flee from every vanity. Let us utterly hate the works of the ways of wickedness.

The body of every human will die. When these people have completed their time of life and the spirit leaves them, their bodies will still die, but their souls will be alive, and they will be taken up.

The soul never returns a second time into the same body of this life. That which has become angelic does not become unrighteous or evil. In this way the soul will not have the opportunity of sinning again by the assumption of the flesh.

In the resurrection, the soul returns to a new body, and both are joined to one another according to their particular nature, adapting themselves, through the composition of each, by a uniform agreement like a building of stones. Hence it appears that the soul is not naturally immortal, but is made immortal by the grace of The Primary One.

All souls are immortal, even the souls of the wicked, for whom it would be better for them, that their souls were vanquished. They are punished with the endless vengeance of a quenchless fire. It is impossible for them to have an end period, so they can be put out of their misery.

The spirit is a subtle, substance, material, which issues and moves forth without form. I believe that all who experience trouble must have committed sins other than what they realize, and so they have been brought to this good end.

But even if a person should happen to suffer without having sinned at all, which is rare, that person's suffering is not caused by the plotting of some angelic power.

A dog is prone to bite and tear the flesh of a person. If the dog should die as a puppy, it is still guilty of sin. The dog is guilty because the sin lives in the dog. The puppy has never bit anyone because it was too young and it hasn't had the occasion to bite someone. If the puppy lived, it would have bitten a person for certain.

It is similar to the suffering of a new-born baby, who seems not to have sinned. A new-born baby has never sinned before. More precisely, it has not actually committed any sins, but within itself, it has the ability of sinning.

Justly so, if by chance a grown man has not sinned by deed and yet he suffers, he suffered the suffering for the same reason as the new-born baby and the puppy has. He has within him, a sinfulness nature, and the only reason he has not sinned was because he has not had the occasion to do so.

If I see a sinless person suffering, despite they have done no wrong, I must call that person evil by nature or by the intent to sin. I will say anything rather than call providence, evil. Nevertheless, let us suppose that you leave aside all these matters and set out to embarrass me. You may bring up the case of so-and-so who must have sinned, since he suffered!

If you permit, I shall say that he did not sin, but was like the new-born baby that suffers. But if you press the argument, I shall say that any human being that you can name, is human. For no one is pure. As someone once said, excellent souls are punished honorably when they die for their religious beliefs.

Other kinds of souls are purified by some other appropriate punishment. We can assume that one part of the so-called will of The Primary One is to love all. A second part of Him is to desire nothing. A third is to hate nothing.

The Primary One said;
Behold, this is the fasting that I want.
Man should humble his soul.
He should lose, every band of iniquity.
Untie the fastenings of harsh agreements.
Restore to liberty, them that are bruised.
Tear into pieces every unjust engagement.
Feed the hungry with your bread.
Clothe the naked when you see him.
Bring the homeless into your house.
Do not despise the humble if you are in their company.
Don't turn away from the members of your own family.

But the way of darkness is crooked, and full of cursing. It is the way of eternal death with punishment.

These are the things that destroy the soul; mental instability, idolatry, over-confidence, the arrogance of power, hypocrisy, double-heartedness, adultery, murder, plunder, haughtiness, transgression, deceit, malice, self-sufficiency, poisoning, magic, and greed.

Also, it is in this way, are those who persecute the good. There are some who hate the truth, and they love falsehood. There are those who don't know the reward of righteousness. Some don't cleave to that which is good. Some mistreat the widow and the orphan.

A Soul Interviewed In Hell

I saw Judas walking with The Primary One, in a punishment of great severity. I said to him, what are you doing in this punishment? Why did the Lord not bring you up with all these souls which He brought up?

Judas said to me, woe is me and double woe for what I did to my Lord. I sinned against Him and betrayed Him for perishable money. I went and took the money and gave it back. I asked Him to forgive me.

I said to Him, will You abandon me for a single thing which I did when I sold You for money? Do not abandon me. Will you look at me as I go to my destruction? Remember, my Lord, that I heard You talking to Peter. When he asked You, if my brother sins against me, how often should I forgive him, up to seven times?

You told him, not just up to seven times but up to seven hundred times. I sinned against You just once. Will You look at me as I go my destruction? My Lord, do not. Who is the man who will watch his son going down to the depth and not help him?

I, even I, who dared to betray You, will You look at me as I go down to my destruction? Do not, my Lord. He then sent me to the desert, telling me not to fear anyone but God. If you see the devil coming, fear neither him nor anyone except God.

I went to the desert in hopes that God might forgive me. The chief authority of destruction, that serpent came to me. He lifted his head above me. His mouth was open and ready to swallow me. I was afraid of him and I worshipped him: You are my lord. He then removed himself from me. I wept.

There is no forgiveness for me. I considered what to do. If the Lord were here right now, I would call Him. He came to this place and he took all the souls and emptied out the souls in hell, except for my soul.

Do Animals Have Souls?

The church in Rome and the Jewish elders expelled the idea that animals have a soul. They were stuck on the idea that man, and not animals were made in the image God.

They felt as though messages from The Primary One should be channeled through them rather than to be channeled through ordinary people. So they rejected the idea that animals had a soul. They wanted superiority among the animals and among men.

As a result of their feelings, they rejected the Gnostic Gospels of Peter, Paul and of James that offered a different belief other than the perverted belief they had. They ignored The Gospel of Mary Magdalene because she was a woman. So the Gnostic Gospels were hidden until now.

Do Animals Have Souls?
These are extracts from The Gnostic Gospels.
The animals that are mentioned here are sacred Deities in various countries and religions.

The Cat and the Dove are especially honored and protected in Egypt. The country of Egypt is the most ancient center of civilization, religion, philosophy and true science.

The cat is not willfully a cruel animal, as it is falsely accused to be cruel by the ignorant. The cat is no more cruel than the baby which torments it. The baby is ignorant of the pain it gives to the cat.

Far more cruel are the human beings, who torture and destroy millions of innocent creatures to gratify a depraved appetite or to satisfy their own lust to kill. Some humans have a lust for cruel experiment.

It is as alleged by both ancient and modern scripts, that the cat is truly the most human of all animals. The cat is the most loving, gentle and graceful of all animals, rather than the more self assertive dog, which is taught by man to hunt and to kill. These other creatures are weak, but the dogs are strong.

Most people lack wisdom and lack love. Understand that every creature, in which The Primary One has made, has its end, and its purpose. Who can say what good is there in that creature or what value that creature has to you, or to mankind?

The cat was an ancient symbol of Deity, because of its ability to see in the dark and its other attributes. More than one recording is given of the protection of these beautiful animals. Even now in some places, they are unjustly despised and regarded with disfavor.

The Primary One is the Friend of all things that suffered. He cast his protection around these innocent creatures. He taught men and women to do likewise, and to feel for all of the weak and all of the oppressed. This beautiful and much criticized animal was a native of Egypt.

Woe unto the man who receives the mysteries, and falls into sin. Such are worse off than the beasts, which you ignorantly killed. All live by one breath, as the one dies so does the other. A man has no preeminence over a beast, for all go to the same place and all come from the dust and return to the dust together.

As all creatures come forward from the unseen world and into this world, so all creatures will return to the unseen world, and so will they come again until they are purified. Let the bodies of them that depart be committed to the elements around you. They must be buried.

DREAMS

DREAMS

When those who are going through dreams wake up, and they sometimes don't see any meaning of the dreams that they may have had. They regard all these disturbances as incidental, for the dreams mean nothing to the dreamer. They don't respect the dream as anything, nor do they regard the solid facts of the dream itself.

Instead, they purposely forget the facts of the dream. Regarding the knowledge of The Primary One, dreams have a value to the dreamer, just as sunrise has its value. This is the way many have acted. They were ignorant of the future before the dream and they were ignorant of the future after the dream.

With this dream, the person can acquire knowledge. Goodness waits for the man who will return and awaken a second time by avoiding the mistakes of the days to come ahead of him. Blessed is he who has opened his eyes. The Spirit ran after that man and gave him forethought of what the future could be.

The dream set him up on his feet, even though he had not risen yet. The dreams gave the dreamers knowledge of Wisdom and of Life. For the ones who were only interested in material things, did not see The Primary One's warnings and they did not know him.

Who Is The First Devil

Extracts From Devils And Their Master

The devil is a living entity bent on doing evil.
Evil is the pull within a living entity to do something that it knows is wrong.
The first devil is a secondary life form compared to the primary king of motion.

The primary life form or king of motion is the one whose body structure moves faster than visible light. This is the fastest rate of ground space occupancy, without disintegration. The Primary One's structure, scientifically, makes it immortal, all powerful and all understanding.

The devil is subservient to The One Who Is All Powerful and All Understanding based on motion skills. The first devils birth along with the birth of the entire universe known to man, was orchestrated by The One Who Is All Powerful. The first devil hates The One.

The devil, who is the one who is the most adverse, believes that the universe was poorly designed and it needs correction. The one, who is most adverse, is a beloved fallen son of The Primary One. The first devil's nickname is The Accuser.

The Accuser does not recognize he is loved and he wants to rub manure in the face of his father. The Accuser wants to make an army of those who agree with him and his ways. He wants to use this army of "the disgruntled". He wants to show the whole world including his father that he can make the world better.

The accuser wants the world to worship him. He wants his father to worship him. He wants to be praised. He wants to be like the one that is most high. He wants those who pass by, to bow their heads as they go by and say good morning sir.

The Accuser's jealousy runs hot against his father. He wants to smite his father. He leaves a trail of misery and destruction where ever he goes. He knows that there is power in numbers. He constantly searches for new recruits. The chief devil wants to visit every creature himself in order to lead his crusade.

He can not visit every one of them himself because of his natural form, makeup and radiance factor. His radiance factor is close to that of gamma rays and would wipe out most, just by his presence. So, he sends associates of his in his place.

When The Primary One visits a man, he takes his right hand and covers their eyes. He alone knows this science. The Primary One had demoted The Accuser by reducing his span of territory and reducing his ability to adjust the cosmos.

Records suggest that The Accuser still retains his ability of movement almost completely unrestrained. He may actively have access to his father's courtrooms. He also is sometimes called as a witness against some, in these hearings.

He still retains the powers of invisibility, transportation, wit, and manipulation of some earthly and some cosmic matters. He never needs to sleep.
The first devil will tamper with anything or anyone. The first devil thinks, if he can collect many beings to be sympathetic to his cause, he can overthrow his father by reason of numbers and by chaos.

He incites the brutalizing of flesh, sadness, wars, amputations, hunger, divorce, guilt, imprisonment, hate, hurricanes and strange tides. His whole life is a constant binge of doing what is wrong.

He has no intention to limit discomfort. If he had his way, everyone's mother would die on the same day. He loves watching fathers beat their children into disfiguration. He laughs when a man and a woman argue. His idea of a good day is, when a city has corrupt politicians and judges.

He loves bad language. His idea of the perfect Sabbath is for a priest to tell the whole parish to go fuck themselves. This is The One Who Is Most Adverse.

Foreclosure

It is agreed by many that The Primary One's closest, living relative as of the year 2017, was born in Buenos Aires, Argentina. This man was selected by sacred lot. He has somewhat of a questionable behavior and a questionable way of life. His background to say, at the very least is a little cloudy, not to mention, he errors daily.

Never the less, he holds court in the best way he can every day. His position is irrevocable. His name is Jorge Mario Bergoglio and was born December 17, 1936, I'm sure many may differ and many may agree, however, who will come forward and take "The Primary One" to court?

4,000 Pages

This <u>edited thesis</u> was presented to you. One might like to read for themselves, the exact unedited versions, and the more accurate version of these extracts. These books contain 4,000 pages and most of them are written in an Elizabethan dialect.

Here is the list of books, manuscripts and documents in which this manuscript was derived. It took the author of this manuscript, decades to read these books and he produced this watered down version of what the books might represent.

This version that the author has produced may have many flaws in it, however, this manuscript is the only one of its type known to him as he types his way through this document as of this moment and as of this time.

The Book List Is

- The Gospel Of Thomas

- The Chronicles of The visiting Gospa at Medjugorje,
  http://www.medjugorje.org/

- Clement To The Corinthians
  Provided Courtesy of Eternal Word Television Network

In Memory Of

Fonder of The Eternal Word Network
Mother Angelica,
Born Rita Rizzo in Canton, Ohio in 1923.
Died at age 92 on Easter Sunday 2015

Beloved of EWTN
Brother Benedict Joseph Groeschel,.
July 23, 1933 to October 3, 2014)

Book List 2

- The Gospel of Jesus Christ
- The Gospel of Mary Magdalene
- The Apocrypha Of James
- The Gospel Of Philip
- The Act Of Peter
- The Gospel Of Judas
- The Gospel Of Bartholomew
- The Acts of Andrew and Matthias
- The Gospel of Andrew And Paul
- Words of Clemens of Alexandrinus
- Who is Yaldabaoth
- The Epistle Of Barnabas
- The Gospel Of Truth

Book List 3

- Devils and Their Master
Amazon.com Title ID # 8268530
ISBN-13: 978-1986697729
ISBN-10: 198669772X

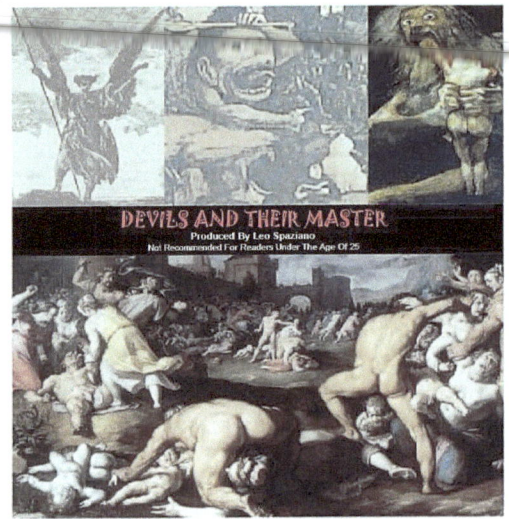

- The Book By Tyndale Please purchase this book on Amazon.com
ISBN 0842332847, 9780842332842

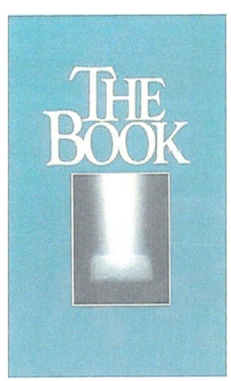

Book List 4

- **The Four Pawed Prophet**
  Amazon.com BOOK NUMBER: 7139545
  ISBN-13: 978-1546451136
  ISBN-10: 1546451137

The Four Pawed Prophet

Good Verses Evil

This Manuscript Describes The Paranormal
And The Science Of Animals, Man And God

Memoirs
Of
A Messenger

Putty

1999 to 2016
Baptized
2002

Page 131

Blessings Of All Faiths
The Primary One's Blessing To You

- In Slovak

Nech je milosť nášho Pána, "Primárny", s vami a so všetkými všade, ktorí sú povolaní Božím skrze neho, kto je pre neho slávu, česť, moc, veľkoleposť a večné nadvládu od večnosti až po večné ,

- In Hinduism

Hamaare bhagavaan kee krpa, "praathamik ek" aapake saath ho, aur har jagah jo ki usake maadhyam se eeshvar kaha jaata hai, usake dvaara usakee mahima, sammaan, shakti, mahima aur shaashvat prabhutv, anant se anant tak rahen. tathaastu.

- Some Western Countries

May the grace of our Lord, "Jesus Christ" be with you, and with all everywhere that are the called of God through Him, by whom be to Him glory, honor, power, majesty, and eternal dominion, from everlasting to everlasting.
Amen.

- In Some Arabic Countries

Qad niemat rabana "alilha" maeakum, wamae kla ma hu matlub min allah min khilaliha, aldhy yakun lah majdi, shurfa, alsiltatu, jalalata, walsiyadat al'abdiatu, min al'abadiat 'iilaa al'abdiat. Amin.

- In some Eastern Countries

May the grace of our Lord, " Jehovah" be with you, and with all everywhere that are the called of God through Him, by whom be to Him glory, honor, power, majesty, and eternal dominion, from everlasting to everlasting. Amen.

- In Some Aramaic Lands, Akal Purakh, meaning timeless One.

"yek'idisiti meḥākeli" ts'ega ke'inanite gari yihuni, inidīhumi be'irisu āmakayineti ye'igizī'ābiḥēri balet'egineti hulu, kekibiru isike zelalemi diresi, le'irisu kibiri, misigana, ḥayili, girimana zelalemawī gizati yihuni. .
Āmēni.

- Nirankar, meaning formless One.

سان اوهان "الله" سان، فضل جي پالٽهار جا اسان اي

سڍيو طرفان جي خدا تي جاء هڪڙي هر ء گڏ،

عزت، پاڪ، ڪيس طرفان جي جنهن ٿو، وجي

كان هميشه دائمي بادشاهي، دائمي ء عظمت.

آمين.

The
End
~

DEDICATION AND CREDIT PAGE

This Science / Religious Manuscript are dedicated to;
- To all of those who think they are unloved,
- To all of those who are in the background and have gone unnoticed,
- To all the ones who took the last train running,
- To all of those who put **religion** into script since the beginning of time,
- To all of those that put **science** into script since the beginning of time.

The author takes no credit for photographic art work or the watered down text in this science / religion manuscript. This manuscript was put together for the soul purpose of the enrichment and the prosperity of all of those who walk the earth.

The author doesn't even know you, but he loves you anyway. He loves the bad and the good as the same. He loves you because his Maker demands this from us. This may be hard for you to understand but the author believes; YOU can make the world a better place even if you are in jail.

Peace Too You.
Peace Too All Of You.
May The Promise of Peace Come To You, And From You.

BIOGRAPHY OF THE AUTHOR

The birth place of the author is PROVIDENCE (The Only City Listed in Dictionaries which means, an all powerful cosmos force) He lived there and witnessed demonic activity against the city's residents for more than half of a century. He believes the demonic activity is directly proportional to the cosmic integrity of the landlord's interest and its name.

The Author is a 5 time patent pending holder, an electronic engineer, a physics student and a Vatican Art collector. Although the author has a terrible criminal police record, he has never told a lie, even if it meant incarceration. He has given 50 % of his income to the elderly and to the poor and has an impeccable record for telling the truth.

The author made a science / religious manuscript which is registered in The National Library Of Congress called; Devils And Their Master. He also put into print, The Four Pawed Prophet. He did this in hopes to stimulate his neighbors to be contrite, or suffer the consequences.

The author has also produced the 2 hour, landscape presentation that depicts The National Forests and Wildlife Reserves in Canada, The USA and The Arctic Circle that is called "On Vacation". Policy was free distribution of the movie to broken families and the poor.

Sold By Amazon.com
https://www.amazon.com/On-Vacation/
UPC#: 191091627754   Title # 850080755

Letter Of Thanks

To CreateSpace.com

To CreateSpace.com

The author cares to congratulate and give true thanks to createspace.com for its generosity, understanding and for the free gifts it gives to the community which are;

- Free international standard book numbers. ISBN.
- Free European standard book numbers. ISBN.
- Free publication of books.
- Free help for amateurs in the editing of movies, and clerical books.

YES, May they be richly blessed for the manufacturing of books, movies, and fine art.

Copyright Page

Copyright © 2018 by Leo Spaziano.

Amazon.com BOOK NUMBER:
ISBN: 13: 978-1986696814
ISBN: 10: 1986696812
Title ID: 8268399

Parts of this book **may be reproduced** and transmitted by any means, electronic or mechanical, as this book expresses the wishes of his Maker, "His Royal Highness"

The author excludes any and all video reproduction or movie making with out his permission.

This is a work of non-fiction. Names, characters, places and incidents are exact.

This book was printed in the United States of America for the benefit of those who suffer chronic grief and seek a remedy for psychiatric disability and chronic weeping.

To order additional copies of this book, contact:
Amazon.com BOOK NUMBER:
ISBN: 13: 978-1986696814
ISBN: 10: 1986696812
Title ID: 8268399

Notes

From;
Leo Spaziano
Johnston Rhode Island
XSpazJrX@hotmail.com

Leo Spaziano (signature)

www.ingramcontent.com/pod-product-compliance
Lightning Source LLC
Chambersburg PA
CBHW051148220526
45473CB00003B/699